The FOSSIL FLORA
of the Drywood Formation of
Southwestern Missouri

The FOSSIL FLORA

of the Drywood Formation of

Southwestern Missouri

by Philip W. Basson

University of Missouri Studies Volume XLIV
University of Missouri Press
Columbia • Missouri

Copyright © 1968 by
The Curators of the
University of Missouri
Library of Congress Catalog
Card Number 67-63045
SBN 8262-7516-8

Manufactured in the United States of America

Acknowledgments

I WISH to express my appreciation to Dr. Joseph M. Wood for his continued companionship, encouragement, and criticisms and for his direction of this study.

Other persons who assisted me are: Dr. David B. Dunn, who offered critical suggestions on the body of this paper; Dr. Walter V. Searight of the Division of Geological Survey and Water Resources, Rolla, Missouri, who provided the cropline maps of the Rowe Coal; Dr. Henry E. Bent, Dean of the Graduate School, who provided funds from the National Science Foundation Cooperative Fellowship Fund, #2520–2600, for transportation during this study.

I wish also to express my appreciation for the invaluable support, encouragement, and assistance given by my wife.

P. W. B.

Ithaca, New York
April 3, 1967

Contents

List of Figures	ix
Introduction	1
I. Historical Review of the Literature	2
II. Methods and Materials	8
III. Geologic Information	13
IV. Fossil Content	37
V. Comparison of This Flora with Other Pennsylvanian Floras	110
VI. Conclusions	134
VII. Summary	136
List of References	137
Plates 1–24	
Illustrations of Plant Macrofossils following page 146	

List of Figures

1. Map of a portion of southwestern Missouri   13
2. Portion of the Lowry City Quadrangle, Missouri   14
3. Portion of the Gaines and Clinton South Quadrangles, Missouri   15
4. Portion of the Arnica and Caplinger Mills Quadrangles, Missouri   16
5. Broadhead's section of type locality (NW$\frac{1}{4}$ SE$\frac{1}{4}$ sec. 25, T. 41 N., R. 25 W.) as amended by Winslow (1891, p. 133)   17
6. Broadhead's section of the center of sec. 26, T. 41 N., R. 26 W. (1873, p. 17)   19
7. Section for an outcrop east of Caplinger Mills, Cedar County, Missouri (SE$\frac{1}{4}$ SW$\frac{1}{4}$ sec. 13, T. 35 N., R. 26 W.)   23
8. Section for an outcrop east of Highway J in the SW$\frac{1}{4}$ NW$\frac{1}{4}$ SE$\frac{1}{4}$ sec. 13, T. 35 N., R. 26 W. of Arnica Quadrangle, Missouri   24
9. Generalized stratigraphic column of Pennsylvanian formations (Krebs Subgroup) in southwest Missouri   29

Introduction

THE PURPOSE OF THIS STUDY is to clarify and enlarge upon a previous research that defined the fossil flora of the Lower Coal Measures of Missouri. The earlier study was based almost entirely on materials that had been collected by individuals in Henry County, Missouri, who had forwarded the floral elements to Washington, D. C., for identification. Since such collections usually are made up of better-preserved materials rather than of all the taxa that may be present, it was felt that the previous collections may have contained a bias, and hence the report did not give a complete picture of the flora.

Recent information on the Rowe Coal reveals an extensive area in which the stratum is exposed. It is questionable whether the previous study, based on a very limited area, gave a true representation of the fossil flora found stratigraphically between the Rowe Coal and the Drywood Coal. Therefore, the major goals of this study were to collect plant macrofossils over a larger area to obtain a larger sample and to define the strata from which these fossils were taken. Secondarily, the flora was also to be more carefully identified in light of the numerous studies made on the Pennsylvanian floras since the original work by White (1899). The flora will be compared with floras from other areas in order to determine similarities and to define more clearly the stratigraphic problems involved.

I
Historical Review of the Literature

THE STUDY OF THE FOSSIL FLORAS of Pennsylvanian Age in the United States has been in progress for almost 150 years. Granger, in 1820, presented the first paper dealing with these flora, based on his study of vegetable impressions connected with the Coal Formation of Zanesville, Ohio. Harlan published a paper in 1834 that dealt with fossil plant remains from bituminous coals of Pennsylvania. In 1836 Hildreth reported *On the Bituminous Coal Deposits of the Ohio Valley, With Notes on Fossil Organic Remains*. Lyell and Bunbury (1846) reported their observations on the fossil plants of the coal field of Tuscaloosa, Alabama. The following year, in 1847, Bunbury described fossil plants from a coal field near Richmond, Virginia.

The decade of the 1850's marked the beginnings of an increase in paleobotanical research. Foster (1853) and Newberry (1853) both reported on fossil plants of Pennsylvanian Age from Ohio. Two studies were made by Newberry on the Ohio flora and were published in 1854. Lesquereux, in 1854, published the first of numerous papers dealing with Pennsylvanian Age floras. The papers reported new fossil species in the anthracite and bituminous coal fields of Pennsylvania. Lesquereux (1857) presented a paper entitled *Paleontological Report on the Flora of the Coal Measures of Western Kentucky*. The 1857 paper was followed by another in 1858, in which Lesquereux described new species of fossil plants from the Coal Measures of the United States.

In the decade of the 1860's, contributions were made by only two researchers. H. C. Wood, in the years 1860 and 1866, wrote *Contributions to the Carboniferous Floras of the United States* and *Contribution to the Knowledge of the Floras of the Coal Period in the United States,* respectively. In the years 1860, 1861, 1868, and 1869, Lesquereux published reports on the fossil floras of Arkansas, Kentucky, Illinois, and the Main Sewanee Jackson Coal of Tennessee, respectively.

The years between 1870 and 1880 were the most productive of descriptions of American Pennsylvanian Age fossil floras. In 1870 Lesquereux reported on the fossil plants of the Illinois coal fields. Fossil plants from the Coal Measures of Ohio were described by Newberry in 1873. Andrews (1875) published two papers describing Ohio plant fossils. *Descriptions of New Species of Fossil Plants from Allegheny County, Virginia* by Meek and *Partial List of Coal Plants from the Alabama Fields* by Lesquereux were published in the same year, 1875. In 1876 Lesquereux described species of marine plants from the Carboniferous Measures, in the Seventh Annual Report of the Geological Survey of Indiana. During 1878 and 1879 Lesquereux published two papers dealing with the genus *Cordaites*. The latter year also saw the publication of the first volume of Lesquereux's *Description of the Coal Flora of the Carboniferous Formations in Pennsylvania and Throughout the United States.* Two subsequent volumes appeared in the years 1880 and 1884. Fontaine and I. C. White (1880) described the

Permian or Upper Carboniferous flora of West Virginia and southwestern Pennsylvania. Lesquereux (1887) wrote a paper on the character and distribution of Paleozoic plants. Lesley's three-volume *Dictionary of the Fossils of Pennsylvania and Neighboring States Named in the Reports and Catalogue of the Survey* appeared in 1889 to 1890.

The decade of the 1890's was most significant, in regard to the present study of Missouri's fossil flora. Hambach (1890) listed 92 species of vascular plants occurring in the Coal Measures of Missouri. David White, in 1893, reported on the *Flora of the Outlying Carboniferous Basins of Southwestern Missouri*. In 1897 and in 1899 he reported on fossil plants from the McAlester coal field, Indian Territory, and, in 1899, the United States Geological Survey published his significant contribution, *Fossil Flora of the Lower Coal Measures of Missouri,* in Monograph #37.

The succession of reports by White continued into the early 1900's. In 1900 two papers appeared: *The Stratigraphic Succession of the Fossil Floras of the Pottsville Formation and the Southern Anthracite Coal Fields* and *Relative Ages of Kanawha and Allegheny Series as Indicated by the Fossil Plants.* In 1903, U.S.G.S. Bulletin 211, dealing with the Upper Carboniferous stratigraphy and paleontology of Kansas, was published by Adams, Girty, and White. Here, D. White reported on the fossil plants from the Upper Carboniferous rocks and Permian formations. This was the first publication dealing with a fossil flora of Kansas. Unger gave an account of the various contributions made to our knowledge of Pennsylvanian Age fossil floras in Pennsylvania in 1907. That same year White presented a report on fossil plants from the

Coal Measures of Arkansas. In 1908, Sellards provided a fairly extensive study in his *Fossil Plants of the Upper Paleozoic of Kansas*. White (1908) attempted to correlate the Upper Paleozoic floras of the United States in regard to their succession and range. This publication was followed by two more by White. In 1913 he reported on the fossil flora of West Virginia, and in 1915 he provided some notes on the fossil floras in Missouri. The first detailed study of Pennsylvanian Age fossil plants in Indiana was provided by Jackson (1916).

Few studies were concerned with Pennsylvanian Age fossil plants in the decade of the 1920's. Noé (1923) wrote about a flora of the Western Kentucky coal field. In 1925 he wrote concerning a Pennsylvanian flora of northern Illinois. Round (1926) correlated the coal flora in Henry County, Missouri, and those of the Narragansett Basin of Rhode Island.

Interest in research devoted to Pennsylvanian Age floras revived in the 1930's. Read (1934) described such a flora in his *A Flora of Pottsville Age from the Mosquito Range, Colorado*. In the same year three other significant papers appeared. Arnold (1934) published a preliminary study of the fossil floras of the Michigan Coal Basin. Jongmans and Gothan presented a paper on the floral succession and comparative stratigraphy of the eastern United States, and compared it with that of western Europe. A small section of this paper discussed Pennsylvanian Age fossils. Hendricks and Read (1934) presented a paper dealing with correlations of Pennsylvanian strata in Arkansas and Oklahoma coal fields. In 1936 Darrah presented a paper on the *American Carboniferous Floras*. A concurrent publication, written by Elias, described the character and significance of the Late Paleozoic flora of Garnett, Kansas. Elias, in 1937, dealt with elements of the Stephanian Age flora in the Mid-Continent region of the United States. In the first of several papers, Janssen (1937) published *A Key for the Identification of Plant Impressions from the Middle Pennsylvanian of Illinois*. The following year a dissertation was written by MacKnight on the flora of the Grape Creek Coal at Danville, Illinois. In 1939 the Illinois State Museum published Janssen's *Leaves and Stems from Fossil Forests*. In 1940 Janssen's second paper, *Some Fossil Plant Types of Illinois*, was published by the Illinois State Museum. Read and Merriam (1940) reported an Early Pennsylvanian Age flora from central Oregon. In 1947, Read presented a paper in the *Journal of Geology* dealing with Pennsylvanian floral zones and floral provinces. A significant addition to paleobotanical literature was Arnold's 1949 paper, *Fossil Floras of the Michigan Coal Basin*.

The 1950's were a productive period for paleobotanical studies of Pennsylvanian Age. Stewart (1950) gave his report on the Carr and Daniels collections of fossil plants from Mazon Creek, Illinois. Condit and Miller, in 1951, described concretions from Iowa that were like those collected around the Mazon Creek area. In 1952 Baxter presented the first two papers describing the Coal Age flora of Kansas. Subsequent papers appeared in 1953 and 1954, of which he was coauthor with Roth and Hartman, respectively. The papers dealt with specific genera or species rather than presenting a review of an entire flora, as many previous studies had done. A second study dealing with the Oregon Early Pennsylvanian Age material was reported by Arnold in 1953. In 1954 Wood and Canright evaluated the status of paleobotany in Indiana with reference to Pennsylvanian Age floras. A paper was published by Shutts and Canright in 1955 describing collections of Pennsylvanian plant fossils in Indiana. Mamay and Read made additions to the flora of the Spotted Ridge Formation in central Oregon in 1956. Another publication dealing with the famous Mazon Creek concretionary material was presented by Langford in 1958; the publication was entitled *The Wilmington Coal Flora from a Pennsylvanian Deposit in Will County, Illinois.* Canright (1959) published a report of the fossil plants of Indiana.

Research in the present decade appears to be following in the path set in the preceding. Read and Mamay (1960) presented a paper dealing with the Upper Paleozoic floral zones of the United States. Wood, in 1963, presented a paper entitled *The Stanley Cemetery Flora (Early Pennsylvanian) of Greene County, Indiana.* Also appearing in 1963 was the paper written by Cridland, Morris, and Baxter concerning the Pennsylvanian plants of Kansas and their stratigraphic significance.

II
Methods and Materials

THE SPECIMENS FOR THIS STUDY were collected from sites in Henry and Cedar counties, Missouri. In addition to the sites from which fossils were taken, other localities were examined in Barton, Cedar, Dade, Henry, and Vernon counties, Missouri. The prospective areas or sites were determined by checking the locations of mines or quarries on the general highway maps of the respective counties and quadrangle maps. Another source of likely fossiliferous localities was the previous literature dealing with (1) the Rowe Coal or, more specifically, the Drywood Formation; (2) the fossil floras in Missouri; and (3) other specifically named areas, such as Henry County, Clinton, Deepwater, or Gilkerson's Ford. The collection site east of Caplinger Mills, Missouri, was brought to my attention by Dr. Carl Chapman, Professor of Anthropology, University of Missouri, Columbia.

Dr. W. V. Searight of the Division of Geological Survey and Water Resources of the State of Missouri provided cropline maps for the Rowe Coal, which were carefully followed and the outcrops checked for the overlying Drywood Formation and its associated plant fossils.

As is often the case, more sites were visited than were productive of materials. The site identifications in older publications were of limited value because, in some cases, the localities of previous collections were not accurately identified. In describing the Pitcher shaft, from which large collections of fossil plants were made, Winslow (1891, p. 139) said, "[The] shaft is about a mile and a half southwest of the Kinney slope and is about three miles from Clinton"; the Kinney slope was "about two miles southeast of Clinton." General county highway maps and quadrangle maps used a single symbol to designate limestone quarries, coal shaft mines, or coal strip pits. The two latter mining operations were the preferred sites; however, accessibility to any fossil plant material varied with the condition of the mine. At some mines, the shafts had slumped in, and the dump heaps alongside were considerably overgrown with vegetation, some with elm trees 40 to 50 feet tall.

A few of the sites that were visited in Henry County, Missouri, and that proved to be unproductive for fossils were:

Clinton South Quadrangle, Missouri
SW¼ SE¼ sec. 17, T. 40 N., R. 26 W.
SE¼ SE¼ sec. 3, T. 40 N., R. 26 W.
NW¼ NE¼ sec. 2, T. 40 N., R. 26 W.
SE¼ NW¼ sec. 35, T. 41 N., R. 26 W.
SW¼ NE¼ sec. 11, T. 41 N., R. 26 W.

Gaines Quadrangle, Missouri
SW¼ NE¼ sec. 17, T. 41 N., R. 25 W.
NE¼ SW¼ sec. 17, T. 41 N., R. 25 W.

The Rowe cropline is rather extensive and follows a highly convoluted pattern extending from the northeast corner to the southwest corner of Barton County, where the cropline is most extensive. The cropline also extends into Cedar, Dade, Jasper, and Vernon counties, Missouri, but to a lesser degree than in Barton County. Although containing the Rowe Coal and overlying shales and sandstones, the areas were either completely lacking in plant fossil material or the plant fossils were in such a fragmentary condition that identification was impossible. Rather than follow the entire cropline across the country, the study emphasized those sites where the cropline crossed various types of roads or where it was especially prominent on a hillside. The Rowe Coal cropline maps were followed for two reasons: (1) to ascertain whether the coal was present and the overlying strata contained plant fossils, and (2) to ascertain whether the shales had any fossils in places where the coal was absent.

One site of the Drywood Formation was the outcrop near the bridge across Dry Wood Creek, north of U. S. Highway 54, in the NE¼ SW¼ sec. 5, T. 35 N., R. 32 W., Vernon County, Missouri, as taken from Searight (1959, p. 26). The succession did have a few fragmentary plant materials, which, due to their poor condition, were not included in this study.

Once located, the samples were gathered either from newly exposed faces in the shales above the Rowe Coal or from different lithologic facies also above the Rowe Coal. The sections from which the samples were taken were noted carefully, and measured columnar sections were made. The specimens were collected in a manner to include specimens from the vertical exposure of the entire fossiliferous layer in the sample. This was accomplished by splitting the shale layers and then placing the samples on cardboard trays in a single layer for transport back to the laboratory. The careful placement and handling prevented loss of valuable specimens through abrasion of the carbonaceous compressions. Specimens with a great deal of mud or debris adhering were rinsed under tap water and allowed to air dry.

In the laboratory, a small strip of white paint was applied to a nonfossiliferous area of each specimen. A catalogue number in India ink was applied to each white strip. I had considered giving a letter designation to those pieces of shale that were counterparts; however, in many cases there was more than one specimen on each piece of shale. In some instances the shale pieces were further split, at which time a new number was given to each piece of shale. The method of designation might have caused an increase in the total number of individual plant remains, but this possibility was taken into account when the relative frequency classification—rare, not common, or common—was applied to each genus or species. In conjunction with the catalogue number, a small dot of colored enamel was applied beside the white strip. The colored dot indicated a specific collection site and helped to differentiate between similar specimens from different sites with the same lithology. A catalogue contained the numbered list of specimens and a color key to the collection sites. After the specimens were marked, they were grouped on the basis of morphological similarity so that identification of similar genera and species could be accomplished more rapidly.

Some fossils were difficult to visualize and identify. When a rock split on different planes, it sometimes left part of the matrix overlying the plant material. A small teasing needle was useful in chipping or prying away the overlying material. Another difficulty in identification was the lack of color differentiation between the fossil and the rock matrix. To overcome this, the surface of the specimen was moistened with xylene, and the contrast between the fossil and the matrix was thereby increased. As the specimens were identified, the generic and species epithets were placed in the catalogue next to the corresponding catalogue number for each specimen. If there was more than one genus and species on the specimen, these names were added to their respective specimen number in the catalogue. The best example of each taxon was then set aside to be photographed.

III
Geologic Information

Geology of the Area

THE FLORA UNDER STUDY was obtained from sites located in Henry and Cedar counties, Missouri (Fig. 1). This area is in the west-central to southwest portions of the state and probably represents the southeastern edge of the Missouri Coal Basin. The locations of the collection sites are as follows:

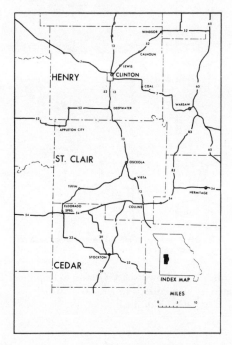

FIG. 1.—Map of a portion of southwestern Missouri. Modified from Geologic Map of Missouri, Missouri Geological Survey, 1961.

Henry County, Missouri

Clary Pit, SW¼ SW¼ sec. 19, T. 40 N., R. 25 W. (Lowry City Quadrangle)

Creek slope, SW¼ SW¼ sec. 26, T. 41 N., R. 26 W. (Clinton South Quadrangle)

Gilkerson's Ford, NW¼ SE¼ sec. 26, T. 41 N., R. 26 W. (Clinton South Quadrangle)

Hilty Mine, NW¼ SE¼ sec. 25, T. 41 N., R. 26 W. (Gaines Quadrangle)

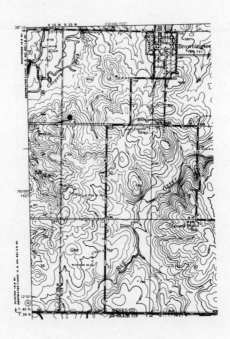

FIG. 2.—Portion of the Lowry City Quadrangle, Missouri. Black dot shows the location of a collection site from which part of the plant fossils for this study were obtained.

Geologic Information 15

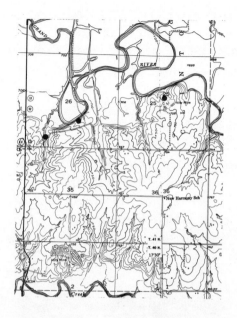

Fig. 3.—Portion of the Gaines and Clinton South quadrangles, Missouri. Black dots show locations of collection sites from which part of the plant fossils for this study were obtained.

Cedar County, Missouri

Harry Good's property, NW¼ SE¼ sec. 2, T. 35 N., R. 26 W. (Caplinger Mills Quadrangle)

Drainage ditch, Route J, SE¼ NW¼ sec. 13, T. 35 N., R. 26 W. (Arnica Quadrangle)

(Figs. 2, 3, and 4). The Clary Pit and the Hilty Mine are strip mines from which the Rowe Coal was taken. The other four sites are areas in which the Drywood Formation is favorably exposed.

In a very early study of the Pennsylvanian strata in Missouri, Broadhead (1873) found that, in addition to coal beds, the Lower Coal Measures in Missouri consist also of sandstone, some beds of iron carbonates, red hematites, and clay iron ore. The sandstone is generally soft, with some shale and clay beds and occasional limestone. In a more detailed manner, Broadhead provided stratigraphic columns for several areas in Henry County, Missouri. He cited the C. B. Jordan mine in sec. 25, T. 41 N., R. 26 W., on the Grand River. The Jordan mine appears to be in the same position as the present-day Hilty mine, from which plant fossils for this study were obtained. Broadhead (1873, p. 16) described the section at the Jordan mine as being 15 feet of blue calcareous slate overlying 32 inches of coal, which in turn overlies 9 feet of blue shale with iron ore concretions. A sandy clay underlies

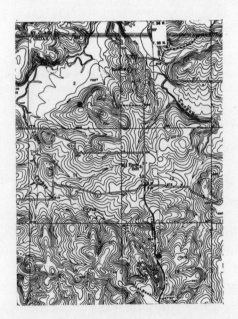

Fig. 4.—Portion of the Arnica and Caplinger Mills quadrangles, Missouri. Black dots show locations of collection sites from which part of the plant fossils used for this study were obtained.

Geologic Information 17

Fig. 5.—Broadhead's section of type locality (NW1/4 SE1/4 sec. 25, T. 41 N., R. 25 W.) as amended by Winslow (1891, p. 133). (Vertical scale distorted.)

the blue shale (Fig. 5). Broadhead compared other mines in the general area to the Jordan mine. No formal proposal of the names *Jordan Coal* or *Jordan bed* appear in this review of the literature, but subsequent writings do refer to the Jordan Coal, so it may be assumed that the Jordan bank was the type locality based on Broadhead's original description of the area. It should be noted that the Jordan Coal and the Rowe Coal refer to identical beds. The difference here is one of time period of usage. Writers from 1873 to 1937 used the term *Jordan Coal;* subsequent to 1937, the term *Rowe Coal* has been used.

Winslow (1891) described the distribution, topography, lithology, and stratigraphy of the Coal Measure strata of Missouri. In the shales overlying the coals, sporadic leaf impressions were found, but the localities were few where such fossils were abundant. The structure of the Pennsylvanian rocks in Missouri was described by Winslow (1891, p. 22) in the following manner, " . . . arranged in a series of strata which have, generally, a slight undulating, westerly dip." Although the coals were most abundant and thickest over the marginal portion of the Pennsylvanian outcrop, the coals were more irregular in character and distribution here than in the interior portion of the basin. Winslow (1891, p. 24) stated the following:

> *The most noticeable features of the stratigraphy of these Coal Measures is the variability of details. Strata are characteristically non-persistent, both as regards thickness as well as material. Beds of coal thin out and disappear; beds of shale pass into sandstone or grade into limestone.*

Winslow (1891, p. 133) said that the mining operations in Henry County, Missouri, were limited to two coal beds: the Hydraulic Limestone bed (Tebo) and the Jordan bed. However, he did not relate the two beds or define their areas. Numerous plant remains and the discovery of a large number of fine specimens by Dr. J. H. Britts of Clinton, Missouri, at the Kinney slope, 2 miles southeast of Clinton, were mentioned by Winslow (1891, p. 139). These specimens were distributed to many prominent paleontologists in the country. It should be noted here that Dr. Britts had an operating coal mine on his own land northwest of Clinton, but the coal taken was from the Tebo coal bed.

Most of the botanical material used by White (1897,

p. 288) had been obtained from several localities in the vicinity of Clinton, from what was known as the *Jordan Coal*. The remainder of his material came from a second, higher seam, 45 feet above the Jordan Coal, located on the Grand River near Gilkerson's Ford. A section by Broadhead (1873, p. 17) was devoted to the Gilkerson's Ford site, "three quarters of a mile west of the Jordan mine, in about the center of sec. 26, T. 41 N., R. 26 W." The following section (Fig. 6) was made for the outcrop about 150 yards upstream (west) from the ford and is the same location from which I collected materials for this study.

The Cherokee Group in Henry County was separated into two divisions by Marbut (1898). The upper

FIG. 6.—Broadhead's section of the center of sec. 26, T. 41 N., R. 26 W. (1873, p. 17), which is about 150 yards upstream from Gilkerson's Ford. (Vertical scale distorted.)

beds, extending from the base of the Tebo Coal to the top of the Cherokee Group, were considered to be consistent enough in character and distribution to be correlated with reasonable certainty across Missouri and into Kansas and Iowa. However, such consistency is not the case with the Lower Cherokee. Hinds (1912, p. 179) made an attempt to correlate some of the Lower Cherokee beds. He said:

> *The Jordan coal lies in places very close to the base of the Coal Measures. It is possible that the Mammoth bed and the coal of the Bowen Trough are contemporaneous with the Jordan, though it is more likely that these troughs and basins of comparatively thick coal are distributed irregularly through the lower part of the Cherokee formation.*

The Mammoth Coal bed was mined at Lewis, and the Bowen Trough near Windsor. Both of the locations are in Henry County, Missouri. Subsequent geological studies have been done in these areas; however, the relationships between the coal beds have not been demonstrated. Furthermore, the paucity of fossil plant material in this area limits opportunity to shed any light on the problem. A possible reason for the dearth of correlation studies is suggested by Hinds and Greene (1915, p. 40). According to these writers, the Lower Cherokee beds were laid down on the irregular surface of older rocks in more or less disconnected bays and estuaries. This accounts for the regional discontinuity of the strata and is the reason that contemporaneous beds vary greatly from place to place. The depth of the Jordan Coal was set at 80 to 100 feet below the Bevier Coal horizon by Hinds and Greene (1915, p. 48), but no direct connection was made between the Jordan Coal and the more persistent strata of the Upper Cherokee. The Jordan Coal is important because the shales above it have been known for a long time to be plant-bearing strata; moreover, White included it in his correlation studies of American and European floras.

Hinds (1912, p. 8) further added to our knowledge of the structure of Pennsylvanian strata in Missouri. He asserted that the rocks are nearly horizontal over large areas, but south of the Missouri River the strata dip in general northwest. He found the dip from Clinton to Kansas City to be 6.4 feet per mile.

One of the factors stressed by Hinds (1912, p. 9) is the remarkable regional persistency in many of the coal beds. Persistency does not refer to their variation in thickness and character, but rather to the fact that coals are more reliable in regional correlation than shales or sandstones. Hinds (1912, p. 11) qualified the above judgment by stating that persistency of coal beds older than the Tebo is less than of those younger and therefore stratigraphically higher beds. The discontinuity in the older beds was further emphasized by Hinds (1912, p. 12) in the following statement:

> *Near Deepwater and Clinton are irregular basins of coal at the Jordan horizon, probably 85 to 120 feet below the Bevier and about 200 feet below the top of the Cherokee. Coal 50 feet or less from the base of the Pennsylvanian and apparently at or near the Jordan horizon crops out in many places in St. Clair and Cedar Counties.*

The strata in the Deepwater area, Hinds (1912, p. 187) concluded, are so variable that detailed sections are of little value. Beds similar to the Jordan Coal were worked in the northwest corner of Dade County and the northeast corner of Barton County. At the time the studies were made, the correlations between the coal in these counties and the Jordan Coal were not known.

In St. Clair County, the coal mined came from two seams, the Tebo, mined near Appleton City, and a bed that lay 100 feet or more lower. This lower coal had many of the characteristics of the coal of Henry County and may be the equivalent of the Jordan. Based on a study of the shales above the coal in this area, Hinds (1912, p. 380) described the following section:

> *The lower coal bed is generally overlain by a few feet of shale, which in turn is overlain with sandstone, but the latter is locally replaced by sandy, argillaceous, or sandy calcareous shale resembling limestone.*

He mentioned no plant fossils in the stratum.

The first delimitation of the Jordan field was made by Hinds (1912, p. 27). He asserted that the field occupied the south-central part of Henry County and much of northwest St. Clair County. A more detailed definition of the area covered by the Jordan Coal in Henry County is on page 179. The Jordan Coal basins were limited to an area bounded by a line drawn from the southwest corner of T. 40 N., R. 26 W., northeast to Clinton, then east to Alberta, and southwest through Brownington to the Henry County line. The areal extent of the Jordan field as shown by Hinds (1912, pp. 28, 29, Pl. III) has been amended by Dr. W. V. Searight (personal communication) to show that the area is of greater extent than Hinds had supposed.

The Pennsylvanian strata in Cedar County, Missouri, have been shown to consist mostly of sandstones of a reddish, brown, buff, or orange color. Based on the 1912 publication by Hinds (p. 135), the following section may be cited for an outcrop east of Caplinger Mills, Cedar County, in the SE¼ SW¼ sec. 13, T. 35 N., R. 26 W. (Fig. 7). In the NE¼ SW¼ sec. 13, T. 35 N., R. 26 W., the coal was found to be 36 inches thick. On the east side of Route J, Cedar County, Millman (1954, p. 109) found the following section (Fig. 8). These measured sections are very close to the present collection site on the east side of Route J in the SE¼ NW¼ sec. 13, T. 35 N., R. 26 W. At this

Fig. 7.—Section for an outcrop east of Caplinger Mills, Cedar County, Missouri (SE¼ SW¼ sec. 13, T. 35 N., R. 26 W.). Modified from Hinds (1912, p. 135). (Vertical scale distorted.)

Fig. 8.—Section for an outcrop east of Highway J in the SW¼ NW¼ SE¼ sec. 13, T. 35 N., R. 26 W. of Arnica Quadrangle, Missouri. Modified from Millman (1954, p. 109). (Vertical scale distorted.)

site, fossil plant material was found in a conglomeratic sandstone, which was not more than 6 inches thick. The conglomeratic sandstone is probably the same as that described by Millman (1954, p. 95) and considered by him to be Upper Cherokee. The "blister" conglomerate included a compact and resistant quartz sandstone matrix. Millman concluded that the pitted or blistery appearance was due to weathering out of elliptical shale pebbles that had been incorporated in the sandstone matrix. The pits were lined and sometimes partially filled with limonite. In this study, the conglomeratic sandstone overlay an unfossiliferous gray shale, which was underlain by a whitish sandstone. The outcrop appeared in a very gently sloping drainage ditch. The thickness of the beds surrounding the conglomeratic sandstone was not determined.

Another question of contemporaneity to the Drywood Formation was raised by Greene and Pond (1926, p. 35) in their study of Vernon County, Missouri. The Dederick Shale was proposed by Greene and Pond for the lowermost shales in the Cherokee, where it crops out in Vernon County. They stated (1926, p. 43) that "The Dederick shale appears to be contemporaneous in age with that over the Jordan coal of Henry County which has furnished many species of fossil plants." The subsequent study by T. K. Searight (1952) has shown this correlation to be incorrect.

In his study of the Humansville Quadrangle, Missouri, which is just east of the Arnica Quadrangle, T. K. Searight (1952, p. 58) reported that the plant remains taken from the Dederick Subgroup were so fragmented that identification was impossible. Thin-bedded sandstones of the Graydon Formation were also found to contain fragmented plant materials. The term *Riverton* supersedes the name *Dederick Formation* in Missouri. The Warner, raised to formational rank by W. Searight and others (1953), replaces the term *Graydon Conglomerate,* includes the beds above the Riverton Coal, and extends upward to the top of the Neutral Coal. Because the present-day stratigraphic column places the Riverton Formation and the Warner Formation below the Rowe Coal and therefore below the Drywood Formation, it must be assumed that no fossils representing the Drywood Formation were present.

In 1937 Pierce and Courtier made a study of some of the strata in eastern Crawford County, Kansas. At that time, they suggested the name *Rowe* for a coal cropping out near the Rowe School. This coal was found to be of Cherokee Age.

Much of the problem with earlier stratigraphic studies was due to the fact that each state, and in some cases local areas, used a different name for the same formation where it outcropped in their respective areas. W. Searight and others (1953), who were representatives of five northern Mid-Continent states, met at Nevada, Missouri, for a conference on classification and nomenclature of rocks of Cherokee Age. The conferees reached complete agreement on division, classification, and nomenclature of pre-Marmaton Desmoinesian beds. The nomenclature retained, with some redefinition, the older established names. The geological surveys of the respective states that participated in the conference recognized the cyclic sedimentation in Cherokee rocks as the basis for formational division. They recognized also that the most persistent units in the succession were the coal beds. Therefore, the divisions of formational rank were made to include the top of one coal to the top of the next higher coal bed.

In his 1954 study, Millman divided the Cherokee into three groups. The lower beds consisted of, from bottom to top, sandy gray shale, calcareous sandstone, and dark gray shale. The middle beds were composed of massive brownish-yellow and brownish-red micaceous quartz sandstone and gray shale. The upper beds consisted of all the strata above the Middle Cherokee Sandstone. In some areas of the Caplinger Mills Quadrangle area, Millman found the Cherokee strata were directly deposited upon Burlington Limestone (Mississippian System), and in other areas the Cherokee strata were deposited upon the Jefferson City Formation (Ordovician System). These findings give further support to the concept of an extensive period of nondeposition or erosion prior to Pennsylvanian deposition.

One of the formations established after the 1953 conference was the Drywood Formation. In his description of the Drywood Formation, Howe (1956, p. 38) noted that it was variable in thickness and lithology. In upward order, the Drywood Formation consists of a basal dark shale, containing lenticular limestone; thin irregular silty limestone and clayironstone; underclay; and the Drywood Coal. The type locality is a tributary to Dry Wood Creek in northwestern Barton County, Missouri. An interesting variation of the above description is included in Carver's (1959, p. 40) and Walker's (1961, p. 99) theses. They both stated that Howe (1956) and W. Searight (1955) described the formation as follows:

> *consisting of a fossiliferous, carbonaceous, black limestone, only locally expressed; dark gray to black fissile shale; gray to brown siltstone and shale with more or less calcareous material; silty underclay; and a coal of variable thickness.*

Carver (1959, p. 40) also found that the oldest Cherokee formation exposed in the Clinton North Quadrangle was the Drywood Formation. He stated the following: "The Rowe coal and the basal part of the Drywood formation are not exposed." I propose that the lack of exposure of the lower strata may be explained by the northwest dip of the strata. A generalized section for the Drywood Formation in the Clinton North Quadrangle, made up from information supplied by Carver (1959, p. 40), revealed a bottom phosphatic shale with a maximum thickness of 4 feet. It is usually black to dark brown and grades into overlying strata. A dark gray argillaceous, fossiliferous, weathered limestone up to 2 feet thick overlies the shale. Carver (1959, p. 41) stated: "Above the phosphatic shale lies an irregular succession of interbedded siltstones and stigmarian sandstones." Both contain abundant carbonized plant fragments. Carver (1959, p. 42) found that the underclay of the Drywood Coal averages 4 feet in thickness. The underclay, which is light to dark gray, also contains abundant fragments of carbonized plant material. The Drywood Coal was found to vary in thickness up to 0.6 feet. As can be known from the above descriptions, no identifiable plant fossils were found in this area, although it closely adjoins the collection sites. It is interesting to note that, although stigmarian material was found in the Clinton North Quadrangle, none was found in the sites from which plant fossils were obtained for this investigation.

Walker (1961), working in the Filley Quadrangle, Missouri, which is adjacent to and west of my Caplinger Mills sites, found good sections of the Rowe and Drywood formations. Walker (1961, p. 99) stated:

The Drywood formation within the Filley Quadrangle consists of claystone or siltstone, shale underclay,

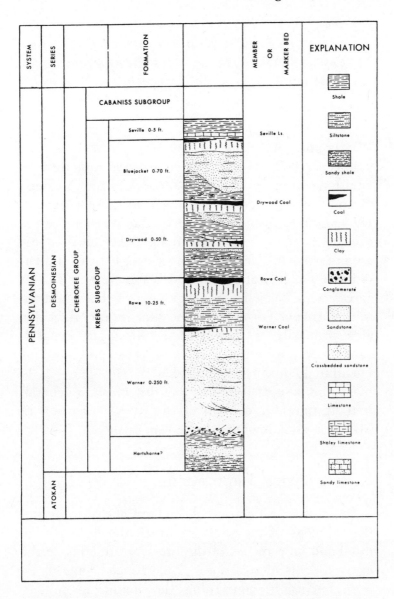

Fig. 9.—Generalized stratigraphic column of Pennsylvanian formations (Krebs Subgroup) in southwest Missouri. Modified from *The Stratigraphic Succession in Missouri*, 1961, Geol. Surv. and Water Resources, Rolla.

> and coal. The claystone consists of a micaceous, dark gray to gray and buff variety with orange limonitic claystone layers. The shale is a gray to light brown fissile unit. A gray, 2 inch underclay is present under the coal. The coal is usually a poor, sub-bituminous variety which is about 6 inches thick.

Walker noted no fossils.

Baker (1962, p. 63) located isolated outcrops of sandstone in the Stockton Quadrangle, which is south of and adjacent to the Caplinger Mills Quadrangle. He tentatively assigned them to the Warner Formation of the Lower Pennsylvanian. Baker checked his assignment with Dr. T. R. Beveridge, then of the Division of Geological Survey and Water Resources, who agreed with Baker in his assignment of this sandstone to the Warner Formation. The Warner occurs in this area primarily as channel fills. No other Pennsylvanian rocks were found in the areas of the Stockton Quadrangle investigated by Baker.

On the basis of the foregoing discussion, a generalized stratigraphic column of the Krebs Subgroup in Missouri, as it is presently known, is shown (Fig. 9).

Paleoecological Implications

Based on consideration of the flora of this study and the lithology of the strata in which the flora was located and on a review of the literature relating to the environments of coal floras, I have made some tentative conclusions concerning the conditions present during the period of coal deposition and subsequent deposition of fossil plant material in a variety of sedimentary rocks in the collection area.

In an early paper dealing with Missouri's coals, Winslow (1891, p. 25) proposed several conditions that, he hypothesized, must have been present during the process of deposition of the Missouri strata. The first condition, which is considered indicative of an environment marginal to a continent, was brackish water. Winslow considered that the brackish condition was favorable for coal bed formation. The second condition was that "the strata, during the process of deposition, must have been at intervals, at or near the water surface, permitting the growth of the coal flora and the accumulation of coal." Since the coal was not accumulated at the time, a possible interpretation of the preceding quotation is that the plant or organic materials accumulated and, under the proper conditions of pressure and time, were converted into coal. Winslow's third condition was that the coal beds were all deposited in basins of varying area and depth. The basins represented great lagoons or swamps in which the organic material accumulated. He explained the lenticular nature of the beds on the basis of the gentle sloping nature of the margin of these basins or swamps. After several attempts to diagram the depositional changes and conditions present during formation of the Missouri Coal measures, Winslow (1891, p. 31) stated the following:

It is of course, impossible to represent in any such diagram the infinitely complex association and the varied succession of strata which resulted from all the combinations of conditions which probably prevailed during the deposition of the Missouri Coal Measures.

Hinds (1912, p. 3) presented additional information on the geological conditions. He wrote that exposure to atmospheric erosion for an extended period caused reduction of the Mississippian surface almost to base level. The sediments of such a low-lying shelf area were seen by Dunbar and Rodgers (1957, p. 316) to be especially susceptible to minor shifts in sea level. This phenomenon is recorded in the Pennsylvanian cyclothems. White (1897, p. 288) contended that the lowest coal-bearing shales, lying in erosion pools, estuaries, or ponds, were deposited by filling or leveling action along the shores of the encroaching Pennsylvanian sea. Therefore, based on the above evidence, the following conclusions may be drawn: (1) the Cherokee strata, more specifically the Rowe Coal and the overlying Drywood Formation, are viewed as typical marginal beds; (2) they were formed in estuary-like basins; and (3) because of this situation they are of limited distribution.

A study of the lithologies of the depositional basins of the Drywood Formation reveals a variety of rock types. The most frequently occurring rock type is shale. Shale deposits, which contain an abundance of woody plant materials, are indicative of deltaic deposits and are similar to the Mississippi delta of today (Shepard, 1959, p. 145). In a paper dealing with marine sediments, Shepard pointed out that shales may be deposited under a variety of conditions. If the deposition was rapid, the stratification was preserved. It was preserved also if deposition took place in bays or in deep holes along the continental shelf, where there were stagnant areas in the water mass. The fine sediments were deposited only where the currents were weak.

Another predominant rock type in the Drywood Formation is sandstone. Sandstone deposition takes place under conditions that suggest a slowing of a rapidly moving body of water. This allows the heavier and larger particles to be dropped. Carver (1959, p. 31) wrote of the variable types of Cherokee sandstones found in Filley Quadrangle, Missouri. He stated: ". . . the sandstones contain only plant fragments as fossils and may be of deltaic or continental origin." In addition, he asserted that the presence of fossilized wood, which he did not define as compression, petrifaction, etc., could not be taken as definite evidence for a fresh-water origin, since fossilized wood could also be expected in rocks deposited near shore along a swampy coast. This theory gained further support from Dunbar and Rodgers (1957, p. 63), who suggested that nonmarine material, such as logs, drift to sea and become buried in marine sediments. Leaves, however, are not likely to survive such transportation, and the more complete specimens are usually buried close to their origin. White (1911) concluded from the relative abundance and condition of the plant fossils in marine deposits that they were good indicators of distance from origin. A finding that I noted and that further substantiates fresh-water deposition was the complete absence of marine fossils. Thus, it may be suggested that the plant fossils were contained in shales and sandstones, which were deposited under fresh-water conditions.

A precaution in interpreting studies of past environments and plants comes from the researchers' inability to determine past environments on the basis of the present flora. The present flora and its structure, because of subsequent evolutionary changes and adaptations, cannot reflect precisely the past environment in which previous plants lived. As Arnold (1947, p. 390) pointed out, the inferences drawn concerning the probable environments of ancient times and based on structural similarities to today's living plants found in certain environments must be tempered.

An example of misinterpretation is illustrated by Arnold's refutation of conclusions drawn by Potonié in his comparison of the relationships between fossil remains, present floras, and past climatic conditions. It was pointed out by Arnold (1947, p. 394) that Potonié advocated a tropical environment for deposition of Pennsylvanian coals. Potonié believed that the abundance of fernlike plants was to be compared to the prevalence of ferns in Recent tropical forests. Potonié reinforced his theory by noting the absence of annual rings in trees of Pennsylvanian Age. Arnold argued that the presence of fernlike plants was not reliable, since the relationships between the Recent and Pennsylvanian species are not known. His strongest argument against a tropical climate was that temperate or cool climates during present times furnish the most suitable environment for peat accumulation. Arnold conceded that, although peat is formed under tropical conditions, it is formed in larger quantities under temperate conditions. Arnold (1947, p. 391) further stated that woods from the Pennsylvanian showed a uniformity of growth and an absence of annual rings. He suggested:

> *Most Paleozoic leaves of which the structure is known would apparently survive very well under modern north temperate zone conditions such as prevail in the eastern or southern portions of the United States.*

The lush plant growth suggests that the climates must have been moist or rainy.

Other uses of plants as indicators of environment

were proposed by Wood (1963, p. 14). Calamite, lepidodendrid, and sigillarian fossils suggest a swampy environment, whereas seeds suggest an upland environment. Ferns as well as seeds are evidences of the presence of a nearby upland environment. The fern fronds or materials should be well preserved to indicate proximity to an upland environment. In a well-preserved state, deposited material could not have been transported any great distance or at least have suffered any rough treatment prior to its deposition. Such was the case with the flora taken from the Caplinger Mills site.

As I found a variety of fossil plant types (stems, leaves, and seeds) and a variety of lithologies at the different collection sites, I must conclude that the depositional environment varied at each site and that each was at a different distance from the original site of plant growth. At two sites, the Hilty mine and the drainage ditch beside Route J in Cedar County, Missouri, seeds were found embedded in a sandstone matrix. Two interpretations may be given these seed findings. They may represent a depositional situation, having been deposited with the sandstones as a heavier sediment, or the seeds in the samples may be representative of an upland species in the flora collected in the area in which they originally fell.

I considered the problem of determining the distance between the site of plant growth and the site of subsequent burial. One specimen of *Spiropteris* sp. was found during the present study. This genus of fern represents a circinate frond, that is, the frond is in a tightly coiled position. In an attempt to determine whether the environment had any effect upon the coiled frond subsequent to removal from the parent plant but prior to its burial, I collected samples of circinate fronds of *Pteridium* sp. and placed them in water, on moist soil, and on a dry surface. The experiment continued for a month, during which time the length and number of pinnule pairs visible on each crozier were recorded. Specimens on the moist soil and on the dry surface lost length in the first week; after that time those conditions of the experiment were discontinued. The specimens in water gained an average of 3.25 cm in length and uncoiled, on the average, three pinnule pairs.

From the foregoing results one may theorize that, although fern fronds may have been ripped or torn from the parent plants in the circinate condition, the subsequent immersion in water allowed the frond to uncoil, at least to a slight amount, so that the *Spiropteris* condition would be relatively rare. In the cases in which the *Spiropteris* condition was preserved, it might be postulated that the specimens were deposited close to their original site of growth and were buried quickly, preventing elongation and uncoiling. Thus, examples of *Spiropteris* in the fossil flora possibly indicate a burial close to the original site of growth.

IV
Fossil Content

List of Plants

THE FOSSIL PLANTS LISTED BELOW were obtained from shales, beds of iron carbonate, and sandstones above the Rowe (Jordan) Coal in Henry and Cedar counties, Missouri. Their relative abundance is indicated as follows: r – rare, 5 specimens or less; nc – not common, 5 to 20 specimens; c – common, more than 20 specimens.

Phylum TRACHEOPHYTA
  Subphylum LYCOPSIDA
    Order LEPIDODENDRALES
      Asolanus camptotaenia WOOD (c)
      Aspidiaria sp. PRESL (nc)
      Knorria sp. STERNBERG (r)
      Lepidocystis fraxiniformis (Goepp.) LESQUEREUX (nc)
      Lepidophloios van-ingeni WHITE (c)
      L. sp. STERNBERG (nc)
      Lepidophyllum sp. BRONGNIART (c)
      Lepidostrobophyllum jenneyi (White) BELL (r)
      L. missouriense (White) comb. nov. (nc)
      L. sp. HIRMER (nc)
      Lycopodites sp. BRONGNIART (c)
  Subphylum SPHENOPSIDA
    Order EQUISETALES
      Annularia acicularis (Dawson) WHITE (r)

A. galioides (Lindley and Hutton) KIDSTON (r)
A. sphenophylloides (Zenker) GUTBIER (c)
A. stellata (Schlotheim) WOOD (c)
Asterophyllites equisetiformis (Schlotheim) BRONGNIART (r)
Calamites carinatus STERNBERG (r)
C. cisti BRONGNIART (nc)
C. cruciatus STERNBERG (r)
C. suckowi BRONGNIART (nc)
Calamostachys paniculata WEISS (r)
C. tuberculata (Sternberg) WEISS (r)

Order SPHENOPHYLLALES

Sphenophyllum emarginatum BRONGNIART (nc)
S. fasciculatum (Lesquereux) WHITE (r)
S. longifolium GERMAR (r)

Subphylum PTEROPSIDA

Orders FILICALES and CYCADOFILICALES

"Alethopterids"

Alethopteris davreuxi (Brongniart) GOEPPERT (r)
A. decurrens (Artis) ZEILLER (nc)
A. grandini (Brongniart) GOEPPERT (r)
A. serlii (Brongniart) GOEPPERT (nc)
A. valida BOULAY (r)

"Sphenopterids"

Diplothmema furcatum (Brongniart) STUR (r)
D. obtusiloba (Brongniart) STUR (r)
Eremopteris bilobata WHITE (r)
Sphenopteris mixta SCHIMPER (r)
S. obtusiloba? BRONGNIART (nc)
S. sp. BRONGNIART (r)

"Pecopterids"
Asterotheca Presl sp. ? (nc)
Mariopteris (Pseudopecopteris) decipiens
    (Lesquereux) White (c)
M. speciosa Lesquereux (c)
M. sphenopteroides (Lesquereux) Zeiller (r)
Pecopteris clintoni Lesquereux (nc)
P. dentata Brongniart (nc)
P. pseudovestita White (nc)
"Neuropterids"
Cyclopteris trichomanoides Brongniart (nc)
C. sp. Brongniart (c)
Linopteris gilkersonensis White (nc)
Neuropteris caudata White (nc)
N. elacerata sp. nov. (r)
N. heterophylla Brongniart (c)
N. missouriensis Lesquereux (nc)
N. ovata Hoffman forma *flexuosa* (Sternberg)
    Crookall (nc)
N. scheuchzeri Hoffman (c)
  Miscellaneous Pteridophyll Organs
Spiropteris sp. Schimper (r)
      Cycadofilicalean Seeds
Holcospermum Nathorst sp. ? (r)
Order Cordaitales
  Artisia transversa (Artis) Sternberg (r)
  Cardiocarpus crassus Lesquereux (r)
  C. late-alatum Lesquereux (r)
  C. ovalis Lesquereux (nc)
  Cordaites crassinervis Heer (c)
  C. principalis (Germar) Geinitz (c)
  Dorycordaites palmaeformis (Goeppert)
      Zeiller (c)

Descriptions of the Fossil Plants

ASOLANUS CAMPTOTAENIA Wood

Plate 1, figures 1, 2

1860. *Asolanus camptotaenia* H. C. WOOD, Phila. Acad. Nat. Sci. Proc. v. 12, p. 238, pl. iv, fig. 1.

1866. *Sigillaria monostigma* LESQUEREUX, Rept. Geol. Surv. Illinois, v. 2, p. 449, pl. 42, figs. 1–5.

1869. *Sigillaria camptotaenia* WOOD, Trans. Amer. Phil. Soc., v. 13, p. 342, pl. IX, fig. 3.

1879. *Sigillaria monostigma* LESQUEREUX, Coal Flora, Atlas, p. 15, pl. 73, figs. 3–6, Text, v. 2, 1880; v. 3, p. 793.

1890. *Sigillaria-camptotaenia m o n o s t i g m a* (Lx.) GRAND'EURY, Geol. et pal. bassin houill. Gard, p. 262, pl. ix, figs. 4–7.

1899. *Sigillaria (Asolanus) camptotaenia* (H. C. Wood) WHITE, Fossil Flora of the Lower Coal Measures of Missouri, U.S. Geol. Surv., Mono. 37, p. 230, pl. LXIX; pl. LXX, figs. 1, 3, 4 (not pl. LXI, fig. 1g; pl. LXII, fig. i; pl. LXIV, fig. e).

Description.—Stem fragment exhibiting fine striations that extend diagonally from leaf cushion to leaf cushion. Leaf cushions spirally arranged. Leaf scar rhomboidal, occupying the upper half of the leaf cushion; lateral angles acute, upper and lower edges rounded, the upper more so than the lower. One cicatricule within each leaf scar, lying slightly above a line drawn between the two lateral angles. The crescentic zone described by White is not seen, nor is the V-shaped ligular depression at the upper rounded angle of the bolster above the leaf scar visible.

Remarks.—The condition described above is thought

to be due to the removal of the epidermal layer of a lycopod stem and the subsequent exposure of the outer cortical surface.

The specimens compare favorably with White's illustrations (1899, pl. LXIX; pl. LXX, figs. 1, 3, 4) and with those figured by Bell (1938, pl. 104, figs. 1–3).

Walton (1953, p. 42) stated that the systematic position of *Asolanus* is uncertain. This has been shown by the numerous attempts to place the cortical material within varied generic groups. White (1899, p. 230) placed *Asolanus* Wood in synonymy with *Sigillaria* Brongniart and identified his specimens as *Sigillaria (Asolanus) camptotaenia*. White (1899, p. 232) supported his judgment by the following statement, "foliar cicatrices borne on small bolsters . . . probably in vertical rows, though plainly affecting a spiral arrangement." It was the character of having leaf scars in vertical rows that motivated White to place his specimens in *Sigillaria*. Grand'Eury (1890, p. 262) placed a group of similar decorticated stem types under the group term Sigillariae-camptotaeniae. This group had previously been placed in the genus *Pseudosigillaria* in the Lepidodendrae. For generic use, for the type described by Wood, Grand'Eury employed the name *Sigillaria-camptotaenia monostigma* Lesquereux. White (1899, p. 238) further pointed out, "Potonié rightly points out the propriety of retaining for Wood's genus, amended, the original name *Asolanus*." Janssen (1939, p. 45) believed that *Asolanus* was a distinct genus closely related to *Lepidodendron*.

Lesquereux's figures (1879, pl. LXXIII, figs. 3, 5) of *S. monostigma* compare very well with the material at hand. Jongmans (1913, p. 4) placed *S. monostigma, S.*

camptotaenia, and *Pseudosigillaria monostigma* in synonomy with *Asolanus camptotaenia.*

To further complicate this matter, several specimens were found that bear a strong resemblance to those figured by White (1899, pl. LXX, fig. 2) and assigned to *Sigillaria (Asolanus) sigillarioides* Lesquereux. White's figure is of very poor quality and does not permit a comparison of materials. This species had originally been assigned by Lesquereux (1880, p. 425) to *Lepidophloios.* Jongmans (1930, p. 443) said that White's specimens of *S. (Asolanus) sigillarioides* looked like an *Asolanus.* Jongmans considered Lesquereux's illustration completely worthless and indeterminable. White (1899, p. 239–241) gave a general discussion relating to *S. sigillarioides,* which Lesquereux had identified as *Lepidophloios ? sigillarioides.* White felt that the two species were closely related, but based their differences mainly on the leaf scar, which was more angular above, the entire vertical diameter being much greater and the lower margins much more convex in *S. sigillarioides.* One of the specimens at hand shows a variation in vertical size of some of the leaf scars even on the same stem section. The question is raised whether the specimens that show a variation in the size of the leaf scar and that in some instances have the leaf cushions overlapping the leaf scars might not be the result of unequal compression during deposition and are therefore not valid as different species.

Several specimens at hand show two layers of the cortical tissue. The internal layer (the cortical layer below the periderm) is termed *Aspidiaria* sp. Presl. The periderm layer, exhibiting diagonal rows of leaf scars and the diagonal furrows between the leaf scars, is known as *Asolanus* Wood. The state of preservation and close relationship of the two layers in the present material permit the correlation of these two layers.

ASPIDIARIA sp. Presl

Plate 1, figure 3

1838. *Aspidiaria schlotheimiana* PRESL, in Sternberg, 1838, Versuch einer Geognostischen Botanischen Darstellung der Flora der Vorwelt, pts. 7 and 8, p. 131, pl. 68, fig. 10.

Description.—Lepidodendrid stem exhibiting a certain state of preservation. *Aspidiaria* condition represented by vertically oriented depressions representing leaf traces. Depressions spirally arranged.

Remarks.—The various stages of preservation of partially decayed materials lead to a confusing situation in regard to correlation of plant structures (stem tissue layers in this case). If the stems have lost the epidermal layers, they are assigned to the genus *Bergeria*. If decay has removed part of the cortical tissue or bark region but has not gone all the way into the xylem, the term *Aspidiaria* is applied to the stem material. The *Aspidiaria* condition is represented by cast and mold impressions.

Weiss (1893, pl. 4, fig. 22; pl. 5, figs. 28, 29) called such material *Sigillaria camptotaenia* Wood. It is closely similar in appearance to the specimen collected in the present study. Sections of the present specimen exhibit what we normally assign to *Asolanus camptotaenia;* these are found where the *Aspidiaria* layer has been removed.

Knorria sp. Sternberg
Plate 1, figure 4; Plate 2, figure 1

1825. *Knorria imbricata* STERNBERG, Tentamen, p. xxxvii, pl. 27.

Description.—Stem section revealing blunt, slightly raised prominences that represent the portions through which the vascular traces passed. Small longitudinal trace visible, leading from below the prominence to small raised point in center of the prominence.

Remarks.—The specimen is the result of the removal of the outer stem surface. One of the specimens figured (Plate 1, fig. 4) reveals a concave depression. The lowermost layer is found to be *Asolanus camptotaenia;* the leaf cushions and diagonal striae between the cushions are evident. Partially covering the *Asolanus* impression is a thin carbonaceous layer. Impressed into this carbonaceous layer are semispherical depressions; this specimen is the reverse surface of fig. 1, Plate 2.

Kidston (1895, p. 546) believed,

> Knorria *may arise from* Lepidophloios, Lepidodendron, *and* Bothrodendron, *or even the Clathrate Sigillariae; it is only a condition of imperfect preservation . . . and hence such specimens . . . do not and cannot possess characters necessary for generic determination.*

LEPIDOCYSTIS FRAXINIFORMIS (Goepp.) Lesquereux
Plate 2, figures 2, 3

1848. *Carpolithes fraxiniformis* ? GOEPPERT ET BERG., De fruct. et sem., p. 26, pl. 3, figs. 33, 34.

1858. *Carpolithes fraxiniformis* LESQUEREUX, in Rogers, Geol. of Pennsylvania, p. 877.

1879–80. *Lepidocystis vesicularis* LESQUEREUX, Coal Flora of Pennsylvania, II, p. 457, pl. LXIX, figs. 18–20.

1879–80. *Lepidocystis fraxiniformis* (Goepp.) LESQUEREUX, Coal Flora of Pennsylvania, II, p. 457, pl. LXIX, figs. 21–23.

1899. *Lepidocystis missouriensis* WHITE, Fossil Flora of the Lower Coal Measures of Missouri, U.S. Geol. Surv., Mono. 37, p. 217, pl. LXI, figs. 1b, 2; pl. LXII, fig. b.

Description.—Isolated s a c l i k e structures (sporangia?), 15 mm in length, 8–14 mm wide, of various geometric shapes, rectangular to trapezoidal, corners rounded. On some specimens two rather deep longitudinal grooves run almost the length of sac.

Remarks.—White (1899, p. 217) stated that he found numerous collapsed, vacant sporangia in close association with bracts of *Lepidophyllum (Lepidostrobus) missouriense*. Some of the material at hand compares very well with his figures. He concluded that these sporocysts were specifically indistinguishable from examples of *Lepidocystis fraxiniformis* (Goepp.) Lx. from Cannelton, Pennsylvania. The latter were found attached to bracts of *Lepidophyllum mansfieldi* Lesquereux. Some of the present specimens also compare with Lesquereux's figures (1878, pl. LXIX, figs. 20, 23), which he identified as *Lepidocystis vesicularis* and *L. fraxiniformis*, respectively.

The generic epithet *Carpolithes* has been retained for seeds, while *Lepidocystis* has been erected as a genus for detached sporangia.

The area between the two longitudinal grooves here is interpreted as being the area that was attached to or appressed to the sporophyll. The two lateral wings are interpreted as being sides of the sporangium that have peeled back from the line of rupture. If one may picture *L. vesicularis* as a ruptured sporangium, it is not difficult to picture *L. fraxiniformis* as a sporangium that has not split open.

These specimens at hand are specifically indistinguishable from *L. fraxiniformis* (Goepp.) Lx., as White (1899, p. 217) has pointed out. In my opinion, the rules of priority indicate that *L. fraxiniformis* is the valid name and should be retained.

As Jongmans (1930, p. 405) pointed out in referring to *L. vesicularis,* "the remains are completely worthless, as far as the formation of a judgment permits." (My translation.) It seems that such variability of shape and size is too dependent upon vagaries of preservation to determine specific epithets with accuracy.

LEPIDOPHLOIOS VAN-INGENI White

Plate 2, figure 4

1899. *Lepidophloios van-ingeni* WHITE, Fossil Flora of the Lower Coal Measures of Missouri, U.S. Geol. Surv., Mono. 37, p. 205–210, pl. LVI, figs. 1–8; pl. LVII; pl. LVIII, fig. 1?; pl. LXI, fig. 1c; pl. LXII, fig. f; pl. LXIII, fig. 5.

Description.—Stem covered by scalelike leaf cushions in diagonal rows. Leaf cushions imbricate, transverse length (1.5 cm) greater than the vertical height (0.5 cm). Leaf scar, 1 cm by 0.3 cm, extending over half the width of the leaf cushion. Lateral angles of cushions acute, curving slightly downward, upper angle gently rounded, lower angle sharper, but not enough to be considered keeled. Leaf scar not well enough preserved in most cases to reveal the ligule, parichnos, or vascular bundle scars. In several cases, three cicatricules are seen within the leaf scar, slightly below the center; the two lateral cicatricules punctiform, the central cicatricule subtriangular.

Remarks.—The *Lepidophloios* material in this study was collected from the same area (Gilkerson's Ford, Clinton, Missouri) from which White received some of his specimens and which he identified as *L. van-ingeni* n. sp. White (1899, p. 205) stated that the bolsters exhibit lateral angles that are very acute at the sides, the lower margins being nearly straight or slightly concave; not carinate. Lesquereux had originally labeled some fragments from the Clinton vicinity *L. dilatatus* Lx. White (1899, p. 206) did not agree that the material from Missouri was the same as the figured originals from Pennsylvania and therefore placed the specimens in *L. van-ingeni*. The specimens in this collection compare identically with that figured by White (1899, pl. LVI, figs. 1, 2a, 2b). However, none of the stem specimens collected compare with those figured by White (1899, pl. LVI, fig. 2; pl. LVII). The other figures listed by White are detached bolsters and lepidodendrid leaves, which will be discussed in their respective groups.

One of the major problems encountered in dealing with fragments of stems is the vertical orientation. There has been a great deal of controversy concerning the position of the leaf scar on the bolster. Kidston (1895), in his excellent historical review of the genus and the problems related thereto, pointed out that many investigators believed that the only difference between the genus *Lepidophloios* and the genus *Lomatophloios* rested on the position of the leaf scar.

He (1895, p. 531) stated that "in *Lepidophloios* the leaf cushions are directed downward and in *Lomatophloios* they are directed upwards." Kidston felt that this single difference was far too insignificant a character for the separation of allied species into different genera, even if the character was constant. But he went on to show that this supposed difference does not hold, even specifically. White (1899, p. 202) stated that the scalelike leaf cushions exhibit the leaf cicatrice at or near the summit. Janssen (1939, p. 47) concluded that the leaf scar was normally found at or very near the top of the cushion, but that in older stems the leaf scar may be at the very bottom of the cushion. Arnold (1947, p. 97) noted that the leaf scar is below the middle line.

Jongmans (1930, p. 444) stated that it is difficult to decide whether we are dealing with a truly independent species in *L. van-ingeni*.

If stem specimens showing a decrease in diameter were at hand, we could more correctly discern the position of the leaf scar on each cushion; however, such specimens have not been recovered in this study. Therefore, it is my opinion that the question remains unanswered, but that the specimens, which are identical with those figured by White, should be retained in White's species in the genus *Lepidophloios*.

Lepidophloios sp. Sternberg

1825. *Lepidophloios* STERNBERG, Versuch einer Geognostischen Botanischen Darstellung der Flora der Vorwelt, v. 1, Tentamen, p. xiii.

Description.—Detached bolster? of *Lepidophloios?* stem. Approximately 32 mm wide and 20 mm high.

Convex inflated structure with long incisionlike longitudinal pit extending 7–10 mm from what is considered to be the basal edge.

Remarks.—The specimens at hand are similar in shape to those bolsters described and figured by White (1899, p. 208, 209, pl. LVI, figs. 3–8) and also found at Gilkerson's Ford. White conformed to the generally accepted conclusion that the flatter or more emarginate border of the leaf scar is its upper margin, the small deltoid or subtriangular pit or trace being on the ventral portion of the bolster.

The placement of this leaf bolster with *Lepidophloios van-ingeni* is questionable because of the much larger size of the former. White (1899, p. 209), in his discussion of the leaf bolsters associated with stem fragments of *L. van-ingeni,* gave only one actual size measurement, and that bolster was larger than the specimen at hand. None of the specimens in the present study was found attached to stem fragments, although they were associated with stem material in the same matrix.

LEPIDOPHYLLUM sp. Brongniart

Plate 3, figure 1

1828. *Lepidophyllum* BRONGNIART, Prodrome d'une histoire des végétaux fossiles, p. 87.

Description.—Leaves long, more than 14 cm in length, narrow, 4–5 mm in width, tapering gently to an acute point. Vascularization simple; a single strand is found in the center of leaf.

Remarks.—White (1899, pl. 61, fig. 1g; pl. 62, fig. i; pl. 64, fig. e) attributed lycopod leaves from the same collection area as of this study as being sigillarioid or

Lepidodendrid leaves, but stated they were probably referable to *Sigillaria camptotaenia.* However, White also considered (pl. 58, fig. 1) large lepidodendrid leaf fragments, which were very similar to the above, as being probably referable to *Lepidophloios van-ingeni.*

No *Sigillaria* stems were found during this study. David White reported *Sigillaria* stems from near Clinton, Missouri, but not from Gilkerson's Ford, where his *Sigillaria (Asolanus) camptotaenia* was found. On the other hand, *Lepidophloios* stem fragments were found in the Gilkerson's Ford material.

Arnold (1949, p. 116) stated, "The detached leaves of *Lepidodendron* and *Lepidophloios* are placed in the organ genus *Lepidophyllum* . . . , and those of *Sigillaria* in *Sigillariophyllum* or *Sigillariopsis.*" He further commented that it is usually impossible to distinguish between the leaves of the different stem genera; those of *Sigillaria* are longer than those of *Lepidodendron.* He also noted Graham's study (1935) of the anatomy of the leaves of Carboniferous arborescent lycopods in which Graham proposed to make the single and the double xylem strand the distinguishing character of *Lepidophyllum* and *Sigillariopsis,* respectively. This can be done only with petrifactions.

On the basis of the facts that (1) no definite sigillarian stem material was found during this study in the area in which the lycopod leaves were found, and that (2) *Lepidophloios* stem fragments were found in the same matrix, the lycopod leaves from this study are placed in the organ genus *Lepidophyllum* Brgt.

LEPIDOSTROBOPHYLLUM JENNEYI (White) Bell

Plate 3, figure 2

1828. *Lepidophyllum* BRONGNIART, Prodrome d'une histoire des végétaux fossiles, p. 87.

1897. *Lepidophyllum* sp. WHITE, Geol. Soc. Amer. Bull., v. viii, p. 298, 300.
1899. *Lepidophyllum jenneyi* WHITE, Fossil Flora of the Lower Coal Measures of Missouri, U.S. Geol. Surv., Mono. 37, p. 214, pl. LIX, figs. 1–3; pl. LXIII, fig. 6.
1927. *Lepidostrobophyllum* HIRMER, Handbuch der Paläobotanik, p. 231.
1938. *Lepidostrobophyllum jenneyi* (White) BELL, Fossil Flora of Sydney Coalfield, Nova Scotia, p. 96–97, pl. XCVIII, figs. 2–5.

Description.—Sporophyll approximately 15 mm long and 5 mm wide, lamina 10 mm long; T-shaped pedicel 5 mm long.

Remarks.—Bell (1938, p. 96) described a sporophyll that he attributed to *Lepidostrobophyllum jenneyi* (D. White). He stated that the species probably belonged to *Lepidostrobus radians* Schimper. Abbott (1963, p. 98) placed similar sporophyll specimens in *Lepidostrobus jenneyi* White. Abbott stated that White originally described and figured isolated sporophylls under *Lepidophyllum jenneyi* and that, with the aid of additional specimens, including cone material, he changed the generic epithet to *Lepidostrobus*. The present study could not determine at which point White changed the generic epithet to *Lepidostrobus*. However, White (1899, p. 215) did make the following statement: " . . . there is hardly room for doubt that we have before us the cone of *Lepidostrobus* or *Lepidophyllum jenneyi*." White merely pointed out the relationship between the two, but certainly did not change the specific epithet. Moreover, although White (1899, p. 212–214) devoted a section to *Lepidostrobus*, the species *L. jenneyi* was not discussed under the latter but, rather, under the genus *Lepidophyllum*.

Hirmer (1927, p. 231) erected the organ genus *Lepidostrobophyllum* for isolated sporophylls.

In view of the fact that the specimen here represented is an isolated sporophyll, the generic epithet *Lepidostrobophyllum* is used in the sense that Hirmer used it. The generic epithet *Lepidostrobus* was first used by Brongniart in 1828, so if separated parts are to be united under a common genus indicating their true relationship, the rule of priority would have us use the generic epithet *Lepidostrobus* Brgt. rather than *Lepidostrobophyllum* Hirmer. The fact that the sporophylls are detached I take as evidence that Hirmer's generic epithet should here be applied.

LEPIDOSTROBOPHYLLUM MISSOURIENSE (White) comb. nov.
Plate 3, figure 3

1828. *Lepidophyllum* BRONGNIART, Prodrome d'une histoire des végétaux fossiles, p. 87.

1899. *Lepidophyllum missouriense* WHITE, Fossil Flora of the Lower Coal Measures of Missouri, U.S. Geol. Surv., Mono. 37, p. 216, pl. LVIII, fig. 2; pl. LX, figs. 1–3; pl. LXI, fig. 1a; pl. LXII, fig. a.

1927. *Lepidostrobophyllum* HIRMER, Handbuch der Paläobotanik, p. 231.

Description.—Sporophyll oblong lanceolate, approximately four times longer than wide, 7.5–10 cm by 1.5–2.5 cm, lamina generally broader at or above the middle, slightly contracted at the point of attachment to the pedicel. Lamina acuminately pointed at the apex. Pedicel triangular in face view, length equal to width at point of attachment to lamina.

Remarks.—The sporophylls that are assigned here to Hirmer's genus *Lepidostrobophyllum* were originally placed in *Lepidophyllum* by White (1899, p. 216). Abbott (1963, p. 97) pointed out that the epithet *(Lepidophyllum)* proposed by Brongniart in 1828 for isolated sporophylls was pre-empted by Cassini in 1816 for a living Compositae.

Jongmans (1930, p. 464) stated that White's *Lepidophyllum* showed a great similarity to *L. majus,* with which White compared it. He further noted that White observed it as a sporophyll of *Lepidophloios van-ingeni.* Jongmans stated that in both cases it concerned very large forms that perhaps should be fitted together.

Hirmer's figure (1927, p. 193, fig. 213) of *Lepidostrobophyllum majus* (Brgt.) Hirmer shows a sporophyll whose lamina has more or less straight sides but terminates in an acute point. The present material compares more favorably with the species description given by White (1899, p. 216) for sporophylls collected at the same Missouri site as the present study. It appears to be more nearly correct to place the detached sporophylls in the genus erected by Hirmer (1927, p. 231) for such a category. For this reason, the specimen is referred to Hirmer's genus.

It must be agreed, as Hirmer and Abbott pointed out, that the characteristics of the laminate portions of isolated sporophylls alone do not provide good criteria for generic differentiation. Since the present material was not associated with strobili, it must, therefore, be placed in the isolated sporophyll genus. There is really no doubt in my opinion that Abbott's *Lepi-*

dostrobopsis missouriensis is the same as *Lepidostrobophyllum missouriense*. Abbott (1963, p. 99, 112) made two appropriate comments that should be borne in mind when dealing with such entities as isolated sporophylls: (1) isolated sporophylls should not be classified under a separate genus when identification with a specific cone is positive; (2) the lamina is significant in specific determination, and the pedicel is significant in generic determination.

It may be true that the lamina and pedicel are significant in specific and generic determination, respectively, but the isolated sporophylls in the present study do not provide any identification with a specific cone. Therefore, the isolated sporophylls are referred to Hirmer's genus *Lepidostrobophyllum*.

Hirmer (1927, p. 231) placed *missouriensis* in the genus *Lepidostrobus*. He described the cones as long and oval; the sporophylls up to 13 cm long; the cone heterosporous. Abbott (1963, p. 100) placed White's species in the genus *Lepidostrobopsis,* which she proposed for unisexual (both megasporangiate and microsporangiate) lycopsid strobili.

Andrews (1955, p. 179) recognized the need for a revision of the genus *Lepidophyllum* by stating that a lycopod cone scale should fall in the genus *Lepidostrobophyllum* Hirmer.

LEPIDOSTROBOPHYLLUM sp. Hirmer
Plate 3, figure 4

1828. *Lepidophyllum* BRONGNIART, Prodrome d'une histoire des végétaux fossiles, p. 87.
1927. *Lepidostrobophyllum* HIRMER, Handbuch der Paläobotanik, p. 231.

Description.—Detached pedicel, oblong-lanceolate, 18 mm long, 9 mm wide. Lamina lacking. Midrib less than 1 mm wide.

LYCOPODITES sp. Brongniart
Plate 4, figure 1

Description.—Stem flexuous, 1.0–2.0 mm wide, some at least 11 cm long. Leaf scars spirally arranged, at least one prominent vascular bundle running the length of stem. Leaves linear-lanceolate, about 1.4 cm long, 1.5–2.0 mm wide, widest below the middle, acutely pointed; midrib running entire length of leaf.

Remarks.—The specimens at hand are tentatively referred to the herbaceous lycopod genus of *Lycopodites*. Since the stem structure revealed nothing significant, the leaves were of major importance in assigning these specimens to *Lycopodites*. It is possible that these specimens are in actuality smaller branches of *Bothrodendron*. Arnold (1949, p. 125) suggested that the resemblance between the two is close. Comparison to some of the herbaceous lycopod materials figured by Lesquereux (1880) showed that the size of the leaves of the present material is larger. Lesquereux used the term *obsolete* in describing the venation of the leaves. This appears to be a good description, since only the midrib is distinguishable in the specimens at hand. The material compares favorably in habit with the specimen figured by Lesquereux (1879, pl. LXII, figs. 8, 8a) and labeled *Lepidodendron weltheimianum,* although the leaves are possibly a little more reflexed in the present samples.

Andrews (1955, p. 182) stated that Seward pointed out that Brongniart first applied the epithet *Lycopo-*

dites to a conifer. Since the term has been used so frequently for herbaceous lycopods, the epithet is used here in the latter sense.

Since there is a question whether the material at hand is a matured herbaceous lycopod or a small branch of a larger lycopod, the specimens are questionably referred to the genus *Lycopodites*.

ANNULARIA ACICULARIS (Dawson) White
Plate 4, figure 2

1862. *Asterophyllites acicularis* DAWSON, Geol. Soc. London, Quart. Jour., 18:310, pl. 13, figs. 16a, b.
1871. *Asterophyllites lenta* DAWSON, Geol. Sur. Canada, p. 29, pl. 5, fig. 60.
1878. *Asterophyllites laxa* DAWSON, Acadian Geology, 3d ed., p. 539. London.
1900. *Annularia acicularis* (Dawson) WHITE, United States Geol. Surv. 20th Ann. Rept., 2:898.

Description.—Leaves, linear-lanceolate, disposed in whorls of 10, widest at middle, tapering apically and basally; average length 14–20 mm; average width 1.0 mm. Ratio of width to length approximately 1:15. Median vein approximately one fourth the width of the leaf.

Remarks.—The single specimen found compares favorably with the description given by Abbott (1958, p. 306–308) and shows a good resemblance to the specimens shown by Bell (1944: pl. 58, fig. 5; pl. 60, fig. 6; pl. 65, fig. 2; and pl. 69, fig. 5). As far as I can determine, this is the first recording of this species in Missouri.

ANNULARIA GALIOIDES (Lindley and Hutton) Kidston
Plate 5, figure 1

1832. *Asterophyllites galioides* LINDLEY & HUTTON, The Fossil Flora of Great Britain, London, 1:79, pl. 25, fig. 2.
1879. *Annularia sphenophylloides* va. *minor* LESQUEREUX, Coal Flora of Pennsylvania, Atlas, pl. 3, fig. 13.
1880. *Annularia emersoni* LESQUEREUX, Coal Flora of Pennsylvania, II: 50–51.
1884. *Annularia cuspidata* LESQUEREUX, Coal Flora of Pennsylvania, v. 3, p. 725, pl. 92, figs. 7, 7a.
1891. *Annularia galioides* (Lindley and Hutton) KIDSTON, Royal Physical Soc., Proc., 10:356.

Description.—Leaves 12 in a verticil, 5.0 mm long, 1–1.25 mm wide, possessing rounded apices. Greatest leaf width above the middle, tapering basally.

Remarks.—There does not appear to be as much variation in the leaf size as is found in some of the other species of *Annularia*. This species has not been previously reported from Missouri.

ANNULARIA SPHENOPHYLLOIDES (Zenker) Gutbier
Plate 4, figure 3

1833. *Galium sphenophylloides* ZENKER, Neues Jahrbuch, p. 398–400, pl. 5, figs. 6-9.
1837. *Annularia sphenophylloides* (Zenker) GUTBIER, Isis von Oken, p. 436.

Description.—Leaves number 11–15 per whorl, 7–9 mm long, 1.5–2.5 mm wide. Leaf widest above the middle, tapering basally, spatulate with a mucronate apex. Midvein occupying about one fifth the width of the leaf and appearing to flare out at distal end. Lateral leaves longer than the leaves parallel to axis.

Remarks.—Based on the numerous specimens collected, my judgment is that this species of *Annularia* was one of the major components of the Missouri flora.

ANNULARIA STELLATA (Schlotheim) Wood
Plate 5, figure 2; Plate 6, figure 1

1820. *Casuarinites stellatus* SCHLOTHEIM, Die Petrefactenkunde, Gotham, p. 397.
1841. *Annularia* spp. HITCHCOCK, Geol. Massachusetts, Final Report, v. 2, p. 542, 754, fig. 226; pl. 22, fig. 2; pl. 23, fig. 1.
1860. *Annularia stellata* (Schlotheim) WOOD, Philadelphia Acad. Nat. Sci. Proc., 12:236.

Description.—Leaves 16–34 mm in length and 3.0 mm wide. The number of leaves in many of the verticils incomplete, but vary from 12 to 15. Recurved tip on some of the leaves.

Remarks.—Specimen #372, although here identified as *A. stellata,* displays the symmetrical pattern and the more exact aspects that Abbott (1958, p. 318) attributes to *A. radiata.* However, the possession of mucronate leaves, which are widest at or slightly above the middle, justifies, in my opinion, the identification of the specimen as *A. stellata.*

In comparison, photographs of *A. stellata* in Gothan and Remy (1957, p. 52) show pointed leaves. Abbott does not mention any recurved tips to the leaves in her discussion of *A. stellata,* but the basal verticil, of her illustrated material (1958, p. 381, pl. 41, fig. 58), has a leaf with a recurved tip. Moreover, Elias' illustrations (1931) of *A. stellata* also show leaves with recurved tips. In no case were any hairs distinguishable on the leaves, or were any globular bodies associated with the leaves. These globular bodies were interpreted by Elias to be seeds.

ASTEROPHYLLITES EQUISETIFORMIS (Schlotheim) Brongniart
Plate 6, figure 2

1820. *Casuarinites equisetiformis* SCHLOTHEIM, Petrefactenkunde, Gotham, p. 397.
1828. *Asterophyllites equisetiformis* (Schlotheim) BRONGNIART, Prodrome d'une histoire des végétaux fossiles, p. 159, 176.

Description.—Leaves linear-lanceolate, widest measurement being at middle of leaf. Leaf length about 1.0 cm; width-length ratio approximately 1:10. Angle between leaves and stem axis about 45°–60°. Median vein occupying one fourth width of leaf. Because of the preservation, the number of leaves per whorl cannot be definitely determined. At least eight leaves were found attached to each of two consecutive nodes. Internodal distance approximately 7.0 mm; leaves somewhat longer than internodes.

Remarks.—The one specimen of *Asterophyllites* collected was of such poor quality that only a tentative identification can be made. The specimen is favorably compared to Arnold (1949, pl. 17, fig. 5).

Abbott (1958, p. 299) discussed the relative sizes of the leaves and the angles that the leaves made with the stem, and the fact that the lower leaves on the axis were attached at approximately right angles, whereas the intermediate and upper leaves were attached at successively smaller angles, the smallest angle being approximately 30°. Furthermore, Abbott stated that the upper leaves were also straighter and less curved than the lower leaves. Abbott stated also that the leaves in ascending order become smaller and that the size of the leaf was dependent on whether the leaf was on a large axis or on a branch of the axis; the larger leaves were found on the main axis. Abbott (1958, p. 301) stated:

> Asterophyllites equisetiformis *is usually represented in collections by ultimate branches, with internodes averaging one cm. long and nodes with leaf whorls whose leaves average about 1.2 cm. long.*

Based on the above statement, it is my opinion that the specimen of *Asterophyllites,* assigned to the species *equisetiformis,* is probably a branch or the ultimate section of a main axis.

CALAMITES CARINATUS Sternberg
Plate 6, figure 3

1824. *Calamites carinatus* STERNBERG, Versuch einer Geognostischen Botanischen Darstellung der Flora der Vorwelt, v. 1, pt. 3, p. 36, 39, fig. 1; pt. 4, 1825, Tentamen, p. 27.
1825. *Calamites ramosus* ARTIS, Antedil. Phytol., pl. 2.

Description.—Incomplete specimen with internode distance of 10 cm. Stem 3 cm wide. Ribs parallel and straight, separated by wide furrow. Branch scar large, with large central opening having diameter of 2 cm or more across margin. Ribs inclining toward each other at margin of central opening.

Remarks.—Arnold (1949, p. 181) considered this species to be one of the most common species of *Calamites* in Europe and probably in North America. Kidston and Jongmans (1917, p. 143) stated that *C. ramosus* Artis has been the generally accepted name, but because of the rule of priority the correct name is *C. carinatus* Sternberg. Therefore, *C. ramosus* Artis is considered to be synonymous with *C. carinatus* Sternberg.

The specimens compare most favorably with those shown by Kidston and Jongmans (1917, pl. 66, fig. 5; pl. 105; pl. 106, figs. 2, 3; pl. 107, fig. 1; pl. 108, figs. 1, 2).

CALAMITES CISTI Brongniart
Plate 7, figures 1, 2

1828. *Calamites cisti* BRONGNIART, Prodrome d'une histoire des végétaux fossiles, v. 1, p. 129, pl. 20, figs. 1-5.

Description.—Internode longer than broad (more than 7.3 cm long by 1.4 cm wide). Ribs tapering to a subacute point, alternating at the nodes; upper end of rib exhibiting oval tubercle. Ribs longitudinally striated and separated by a narrow furrow that exhibits two prominent lines at side of furrow.

Remarks.—The specimen compares favorably with the description given by Kidston and Jongmans (1917, p. 123–125) for this species.

CALAMITES CRUCIATUS Sternberg
Plate 8, figure 1

1825. *Calamites cruciatus* STERNBERG, Versuch einer Geognostischen Botanischen Darstellung der

Flora der Vorwelt, v. 1, Fasc. 4, p. 46, Tentamen, p. xxvii, Pl. 49; v. 2, 1833, Fasc. 5, 6, p. 48.
1935. *Calamites cruciatus* Gutbier, Zwickau (pars), p. 18, pl. 2, fig. 15 (not 9, 10, 12, 13, 16).

Description.—Pith cast with internodes longer than wide (8.5–13.5 cm long by 4.4–5.5 cm wide). Ribs, prominent and rounded, alternating and converging at branch scars. Rib ends pointed. Four branch scars per node discernible, but only by careful scrutiny of converging ribs; otherwise two branch scars per node seemingly present. Branch scar spherical, approximately 2 mm in diameter.

Remarks.—The upper portion of the stem at hand shows great compression, which is indicative of a thin woody region.

The specimen compares most favorably with the specimen shown by Kidston and Jongmans (1917, pl. 130, fig. 3) and designated by them without a description or discussion as *C. cruciatus elongatus* Gutbier. The most distinguishing feature of the specimen, the length-width ratio of the internode, which may be due partially to lateral compression, allows the placement of the specimen in the species *C. cruciatus* Sternberg.

Calamites suckowi Brongniart
Plate 7, figure 3

1828. *Calamites suckowi* Brongniart, Prodrome d'une histoire des végétaux fossiles, v. 1, p. 124, pl. 15, figs. 5, 6; pl. 16, fig. 2 (not pl. 14, fig. 6; pl. 15, fig. 1; pl. 16, fig. 1).

Description.—Pith cast showing only a small portion of the stem or rhizome in nodal region. Ribs 4–5 mm wide, flat, at base flat to bluntly rounded. Tubercles at upper ends of ribs large (6 mm by 2 to 3 mm) and oval. Tubercles at bases of ribs smaller (1 mm) and spherical.

Remarks.—The small portions of this specimen fit the description given by Kidston and Jongmans (1917, p. 111) and compare with plate 87, fig. 3. There is a great deal of variability in this species.

CALAMOSTACHYS PANICULATA Weiss
Plate 8, figure 2

1876. *Calamostachys paniculata* WEISS, Steink. Calam. I, p. 59, pl. XIII, fig. L; Calam. II, p. 173, pl. XIX, fig. 3; pl. XXI, fig. 6.

Description.—Cone fragment, elongate. Sterile verticils 5 mm apart. Sporangiophores attached midway between sterile bracts. Bracts narrow, spreading, forming a right angle with axis, arcing upward to extend beyond lower edge of next whorl of bracts. Cone approximately 4–5 mm in diameter from outer edge of sporangiophore to axis.

Remarks.—Jongmans (1911, p. 303) reported that *C. paniculata* was the cone of *Calamites cruciatus.* The leaves on the stem are linear, as are those of *Asterophyllites.*

The material at hand compares well with that figured by Jongmans (1911, p. 303–306).

Wood (1963, p. 44) pointed out that this species had not been found in many American floras; Bell (1944, p. 105) reported the species from the Cumberland Group of Nova Scotia.

CALAMOSTACHYS TUBERCULATA (Sternb.) Weiss
Plate 8, figure 3

1826. *Bruckmannia tuberculata* STERNBERG, Ess. Fl. monde prim., I, fasc. 4, p. 45, pl. XXIX, pl. XLV, fig. 2.

1876. *Stachannularia tuberculata* WEISS, Steink. Calam. I, p. 17, pl. I, figs. 2–5; pl. II, figs. 1–3, 5; pl. III, figs. 3–10, 12.

1884. *Calamostachys tuberculata* (Sternb.) WEISS. Steink. Calam. II, p. 178.

Description.—Cone elongate, bract whorls 4–5 mm apart, spreading upward or slightly recurved and then curving upward to level of next verticil of bracts. Cone axis not clearly discernible. Sporangiphores occupying all of height between succeeding bract verticils, but attached in middle of internode.

Remarks.—There is good agreement between the specimen and those figured by Jongmans (1911, p. 293–294) and described by him as *C. tuberculata*. This species has been found in direct connection with leaves of *Annularia stellata* (Schlotheim) Wood.

Wood (1963, p. 44) reported that other investigators had pointed out that *Annularia stellata* and *C. tuberculata* were borne on stems of *Calamites multiramis* Weiss. I agree with Wood that this assignment presents a problem, since *C. multiramis* was not found in this flora or in the Indiana flora.

SPHENOPHYLLUM EMARGINATUM Brongniart
Plate 6, figure 4

1828. *Sphenophyllum emarginatum* BRONGNIART, Prodrome d'une histoire des végétaux fossiles, p. 68.

1880. *Sphenophyllum schlotheimii* LESQUEREUX, Coal Flora of Pennsylvania, v. 1, p. 52, pl. 2, figs. 6–7.

Description.—Leaves 7–8 mm long, 2.5–3.5 mm wide. Six leaves per whorl, each leaf dichotomized three times. Leaves with obtusely rounded lobes and teeth at distal end. Several specimens with eight leaves per verticil also exhibiting leaf length of 9–10 mm. Leaf width in latter case 4–5 mm.

Remarks.—There are definite size differences between some of the specimens collected and those given by Abbott for the species. The ratio of length to width is greater in several specimens than that for *S. emarginatum*, but is similar to the ratio given for *S. cuneifolium*. However, the obtusely toothed lobes appear to be the diagnostic characteristic. According to Abbott (1958, p. 341), *S. emarginatum* may be easily distinguished from *S. cuneifolium*. The teeth and lobes of the former are obtusely rounded, while those of the latter are acutely pointed. Since the length varies from 5–10 mm while the width varies from 2.5–4.0 mm, the question arises as to whether the smaller specimens might be interpreted as immature types, since the obtuse lobing is the same. In my opinion, the specimens at hand should be assigned to *Sphenophyllum emarginatum* Brongniart.

SPHENOPHYLLUM FASCICULATUM (Lesquereux) White
Plate 8, figure 4

1880. *Asterophyllites fasciculatus* LESQUEREUX, Coal Flora of Pennsylvania, v. 1, p. 41, pl. 3, figs. 1, 2.

1899. *Sphenophyllum fasciculatum* (Lesquereux) WHITE, Fossil Flora of the Lower Coal Measures of Missouri, U.S. Geol. Surv., Mono. 37, p. 183–187, pl. 50, figs. 1–4.

Description.—Leaves borne in whorls. Leaves 3–6 mm long, approximately 1.0 mm wide, widest at proximal end, pointed at distal end, bilobed for over half the length. Stem ribbed, nodal areas swollen, internode length 3–4 mm, width 1–2 mm. Leaves forming angle from 45°–90° with stem, curving upward immediately, leaf apex almost reaching level of next node.

Remarks.—The leaf number cannot be determined in the specimens at hand. Apparently, Lesquereux experienced the same difficulty (1860, p. 310), for he described the leaves of *Asterophyllites gracilis* in the following manner: "The leaves, never flattened, are imbedded in the stone in such a way that the horizontal section shows only their thickness and their direction." He felt that the peculiar disposition of all the whorls, which indicated the leaf hardness, separated it from other species. The present material appears to have the same characteristic.

The present material compares well with specimens figured by Bell (1940, p. 129, pl. X, fig. 3) but assigned by him to *Asterophyllites charaeformis*.

The material at hand also fits within the description given by Abbott (1958, p. 342, 344), but it does not correspond as well to the figures shown by her (pl. 37, figs. 25–28; pl. 48, fig. 84). Abbott's figure 84, a portion of the original material, shows more opposite branching than the present material.

Sphenophyllum fasciculatum may be distinguished

from *A. charaeformis,* with which it may be confused, on the basis of the bilobed leaves and opposite branches, which are present on the former. Abbott (1958, p. 343) referred to Lesquereux's figure (1878, pl. III, fig. 1) of *A. fasciculatus,* which she identified as *S. fasciculatum.* Lesquereux's figure and Abbott's figure (pl. 46, fig. 74), both identified as *S. fasciculatum,* might easily be confused with *A. charaeformis* on the basis of general habit.

White (1899, p. 183) transferred *A. fasciculatus* Lx. to *S. fasciculatum* (Lx.) on the basis of the general aspect of the plant, in which he felt the obvious characters were strongly similar to those of the common Sphenophylla. He did state, " . . . it is only after a glance at the leaves that one pauses to inquire whether it belongs to some other group." I also question the shape of leaves in *Sphenophyllum fasciculatum* and their assignment to that group, since the first impulse was to assign this specimen to *Asterophyllites* on the basis of its leaves.

SPHENOPHYLLUM LONGIFOLIUM Germar

Plate 6, figure 5

1837. *Sphenophyllum longifolius* GERMAR, Isis von Oken, p. 425, pl. 2, figs. 2a, b.
1843. *Sphenophyllum longifolium* (Germar) GEINITZ, Gaea von Sachsen, p. 72.

Description.—Leaves 2–2.5 cm long, 1.0–1.5 cm wide at the distal end. Leaves with veins, which dichotomize four times, and toothed distal margins. Deep central cleft; lateral clefts not so deep. Leaves five in number, but a small basal remnant gives impression of at least one more leaf; thus leaves total six per whorl.

Remarks.—White (1899, p. 181) placed *S. longifolium* in synonymy with *S. majus*. White stated:

> Sphenophyllum majus *is represented in the Lacoe collection in the U. S. National Museum by a number of specimens from the vicinity of Clinton, Missouri, labelled* Sph. longifolium *by Professor Lesquereux.*

Abbott (1958, p. 349–350) noted that *S. majus* may also possess dissections of varying depths. She had previously noted (1958, p. 348) that "lateral clefts may assume various depths but are as deep as the central cleft," in reference to *S. longifolium*. This material compares more favorably with specimens illustrated by Janssen (1939, p. 95, fig. 79e), in which the middle cleft is much deeper than the lateral clefts.

I am of the opinion that the inclusion of *S. longifolium* with *S. majus*, as indicated by White's synonymy, is incorrect and that the two species should be treated as separate entities.

Since there is such prominent clefting and the ratio of leaf length to width is larger than in *S. majus*, the material at hand is assigned to *S. longifolium*.

ALETHOPTERIS DAVREUXI (Brongniart) Goeppert
Plate 10, figure 2

1828. *Pecopteris davreuxi* BRONGNIART, Prodrome d'une histoire des végétaux fossiles, p. 57.

1836. *Alethopteris davreuxi* GOEPPERT, Die Fossiliean Farrenkrauter (Systema Filicum Fossilium), Nova Act Leopoldina, Bd. 12, p. 295.

Description.—Pinnules triangular to oblong, with bluntly rounded apices, 0.7–1.1 cm long, 4–5 mm wide. Widest part of pinnule primarily in its lower third. Pinnules attached more or less obliquely to their rachis, decurrent, and united by about 2 mm of the lamina. Upper pinnule margin sometimes contracted at the base. Terminal pinnule lacking in specimen at hand. Midrib of pinnule clearly visible and continuous almost to apex of pinnule; lateral veins given off 0.5 mm apart on the midrib, arising at an angle of 75°–80° and dividing once. Veins curving out from the midrib only slightly; thereafter, their course rather straight. Subsidiary veins from the midrib also entering pinnule directly from pinna rachis.

Remarks.—The specimen at hand compares well with that described by Crookall (1955, p. 8). The numerous fine lateral veins leaving the pinnule at a greater-than-60° angle help to distinguish *A. davreuxi* from that of *A. valida*. Crookall (1955, p. 10) gave the stratigraphic range of *A. davreuxi* as being from Westphalian B to D.

ALETHOPTERIS DECURRENS (Artis) Zeiller
Plate 12, figure 1

1825. *Filicites decurrens* ARTIS, Antedil. Phyt., pl. xxi.
1886. *Alethopteris decurrens* ZEILLER, Bass. Houill. de Valenciennes, p. 221 (1888); Atlas (1886), pl. xxxiv, figs. 2, 3; pl. xxxv, fig. 1; pl. xxxvi, figs. 3, 4.

Description.—Pinnules linear, oblique to rachis, approximately 2 cm in length, 2–3 mm wide, slightly arching upward. Lower margins of pinnules decurrent. Midrib prominent, lateral veins leaving midrib at almost right angles.

Remarks.—The specimen at hand is so poorly preserved that a photograph is not useful. Crookall (1955,

p. 27) stated *A. decurrens* may be differentiated from *A. lonchitica,* with which it may be confused, on the basis of the linear pinnules and pinnule width (3 mm) of the former, while *A. lonchitica* has slightly swollen pinnules of from 3–5 mm width, which are lanceolate. He placed *A. decurrens* as most frequently found in the Westphalian B, although ranging from Westphalian A to C.

ALETHOPTERIS GRANDINI (Brgt.) Goeppert
Plate 10, figure 3

1832. *Pecopteris grandini* BRONGNIART, Prodrome d'une histoire des végétaux fossiles, v. 1, p. 286–289, pl. xci, figs. 1, 2.

1836. *Alethopteris grandini* GOEPPERT, Die Fossiliean Farrenkrauter (Systema Filicum Fossilium), Nova Acta Leopoldina, Bd. 17. Suppl., p. 299–300, 301.

Description.—Frond fragment showing secondary pinna approximately 5 cm in width. Terminal pinnule lacking, lateral pinnules arranged alternately along the rachis and obliquely attached. Roundly terminated pinnules broad, 8–9 mm wide, 24–26 mm long, usually separated from each other by 1 mm. Margins mostly parallel, lower margin decurrent. Midrib clearly visible, but not continuous to pinnule apex. Lateral veins forming an angle with the midrib of 50°–60° and arching toward the margin, branching once or twice.

Remarks.—Crookall (1955, p. 31) described specimens that he related to *A. grandini* Brongniart P. Bertrand emend., which are of Westphalian Age. The basic features are very similar to the present specimen except for the larger size of the latter. Crookall listed

A. grandini as occurring in Westphalian C and D of England.

ALETHOPTERIS SERLII (Brgt.) Goeppert
Plate 11, figure 1

1828. *Pecopteris serlii* BRONGNIART, Prodrome d'une histoire des végétaux fossiles, p. 57.
1836. *Alethopteris serlii* GOEPPERT, Die Fossiliean Farrenkrauter (Systema Filicum Fossilium), Nova Acta Leopoldina, Bd. 17. Suppl., p. 301, pl. xxi, figs. 6, 7.

Description.—Pinnules oblong, roundly pointed, lying close together or touching by their lateral margins, 13–18 mm long, 6–7 mm wide, lateral margins more or less arched outwards so that the pinnule is slightly enlarged about the middle. Pinnules oblique to the rachis, adjacent pinnules separated from each other by a sharp sinus, lower margin typically decurrent, upper margin slightly contracted at base. Midrib distinct to pinnule apex, slightly decurrent at base. Lateral veins fine, closely placed, straight, remaining simple or dividing once, branching from the midrib at almost a right angle. Subsidiary veins enter the pinnule from the pinna rachis.

Remarks.—In *A. serlii*, the widest part is usually above the middle and has a sharp sinus between adjacent pinnules, whereas in *A. valida* the widest part is in the lower third of the pinnule and the pinnules are separated by a round sinus.

In *A. serlii*, as differentiated from *A. grandini*, pinnules are narrower in proportion to their length, the sharp sinus separates adjacent pinnules, and the midrib continues to the apex.

White (1899, p. 118) erected a variety *missouriensis* for specimens with long, slender, simple pinnules with borders in many specimens folded under. He noted that specimens labeled *A. lonchitica* from Missouri were really *A. serlii* var. *missouriensis* with their borders buried in the matrix.

Crookall (1955, p. 22) listed *A. serlii* as occurring more or less commonly, both in Westphalian C and D.

Alethopteris valida Boulay
Plate 11, figure 2

1876. *Alethopteris valida* BOULAY, Terr. Houill. du Nord de la Fr., p. 35, pl. 1, fig. 8.
1906. *Desmopteris integra* GOTHAN in Potonié, Abbild. u Beschreib. foss. Pflanzen-Reste, Lief. IV, No. 64, figs. 1, 2.
1914. *Alethopteris integra* KIDSTON, South Staffordshire Coal Field, Trans. Roy. Soc., Edin., v. 1, p. 103, text-figs. 3, 4.
1932. *Validopteris integra* (Gothan) P. BERTRAND, Bass. Houill. de la Sarre et de la Lorraine, v. 1, Flore Fossile, 2me. fasc.

Description.—Pinnules terminating bluntly, 15–20 mm long, 4–5 mm wide, bearing broad shallow blunt lobes, oblique to rachis. Midrib well defined close to rachis, becoming indistinct at about two thirds the length from base of pinnule. Lateral veins of pinnule leaving the pinnule midrib at a 45° angle and usually dividing two to three times, the first fork occurring within 1 mm of midrib, the other two randomly between the first fork and the margin of the pinnule, with each lobe of the pinnule receiving its total venation from a single branch of the midrib. Some pinnules on the pinna decurrent, others exhibiting free base.

Remarks.—As Crookall (1955, p. 16) stated, *A. valida* shows some resemblance to *A. integra* Gothan in the entire or slightly lobed structures that occur in the apical portions of the frond. Bell (1938, p. 69) united *Validopteris integra* and *A. valida* because of the intermingling of the two forms of primary pinnae on a frond—one, the entire, connate, alethopterid pinnules; the other, the elongate, lobate pinnules. He also noted the close similarity of *Desmopteris integra* Gothan to *A. valida,* which P. Bertrand had pointed out. The genus *Validopteris* was apparently erected by P. Bertrand (1932, p. 103) to include those characteristics exhibited by the apical region of a frond from a large *Alethopteris,* notably *A. valida* Boulay.

Crookall (1955, p. 17) found *A. valida* to be almost wholly confined to Westphalian A and B.

DIPLOTHMEMA FURCATUM (Brongniart) Stur
Plate 12, figure 2

1829. *Sphenopteris furcata* BRONGNIART, Prodrome d'une histoire des végétaux fossiles, p. 50.
1836. *Hymenophyllites furcatus* GOEPPERT, Syst. fil. foss., p. 259.
1877. *Diplothmema furcatum* (Brongniart) STUR, Culmflora, II. Ostr. Waldenb. Sch., Abh. K. K. Geol. R. A., Wien, VIII, 2, p. 121, 124 (230).
1879. *Eremopteris missouriensis* LESQUEREUX, Coal Flora of Pennsylvania, Atlas, p. 9, pl. liii, figs. 8, 8a; text, v. 1 (1880), p. 295.

Description.—Rachis striated. Pinnules alternate, diverging at an angle of about 45°, ovate or deltoid, basal pinnules palmate, lobes appearing to be linear. Primary vein decurrent, forking near the base, branching again at the base of each lobe, a single vein going to the apex of each lobe. Lamina poorly preserved, but in better preserved areas there appears to be a striated surface.

Remarks.—White (1899, p. 16) felt that there was a progression of closely related varietal types with delicately dissected pinnae ranging from *E. missouriensis* to *Diplothmema furcatum.* He stated that perhaps the Missouri specimens should be placed in the latter genus as restricted by Zeiller. White found that the lower basal pinnules in the lower part of the frond presented a *Sphenopteris spinosa* form, while those near the upper end of the pinnae were a *S. furcata* form. These two forms were found by White (1899, p. 17) to be part of the marginal portion of a frond of *Eremopteris missouriensis.* Further, White was unable to find any distinctive characters by which Missouri specimens, labeled *S. spinosa* Goeppert and *S. splendens* by Lesquereux, could be separated even as a variety from *E. missouriensis.*

Bell (1938, p. 27, pl. X, fig. 3; pl. XI, fig. 2) placed closely comparable specimens under *Sphenopteris missouriensis* ? (Lesquereux) White. I believe this material to be the same as the specimens in this study and completely different from that described by White (1899, p. 43, pl. XIV, figs. 1, 2).

Janssen (1940, p. 56–58) placed *Hymenophyllites splendens* Lx. under *Diplothmema zobeli* Goeppert. He compared his material with Lesquereux's specimen of *Hymenophyllites splendens* and found them to be the same. He also said that his specimen bore a marked similarity to *Eremopteris missouriensis* Lx.

and that they belonged to the same species, since they had no characters by which they could be differentiated. Janssen concluded that *H. splendens* Lx. did not represent a distinct form and referred it to *D. zobeli*. However, his illustration is rather poor and does not allow a judgment to be made on the basis of illustrated material. Jongmans (1960, p. 1140) did not relate *D. zobeli* to *D. furcatum*; however, since Jongmans probably did not examine the specimens, Janssen's conclusion would be more valid.

Jongmans and Dijkstra (1960, p. 1117–1119) discussed the problem of synonymies in *Diplothmema furcatum*, which they found to be a rather extensive list. They noted on p. 1090 that the generic epithet may be spelled *Diplothmema* or *Diplotmema*.

A problem remains, since both White and Janssen were rather positive in their placing *Sphenopteris (Hymenophyllites) splendens* Lx. under their species and since Janssen compared his specimen of *D. zobeli* favorably with White's *Eremopteris missouriensis*, which Jongmans recognized as being under *D. furcatum*. Jongmans and Dijkstra separated *D. zobeli* and *D. furcatum*. It is clear that this group, which exhibits such variability, is in need of additional careful study.

Wood (1963, p. 51) placed comparable material under *Palmatopteris furcata*, pointing to Arnold's (1949, p. 205) statement as a basis for his judgment. It does not seem that White presented a good case for Potonié's genus *Palmatopteris*, as Arnold stated. White (1943, p. 93) wrote the following:

> *Still later in 1893, species that are typical of Stur's original genus and that constitute the nucleus about*

which he organized Diplothmema *and for which that name should have been retained by Zeiller were removed by Potonié, who applied to this group of species the new name* Palmatopteris, P. furcata *being stressed as representative of the new genus. It is, however, to be borne in mind that even as late as 1893 little regard was given by most European paleobotanists to the matter of type species of genera and their retention where possible under the original generic designation.*

Jongmans and Dijkstra (1962, p. 2134–2139) discussed the synonymies associated with *Palmatopteris furcata* (Brongniart) Potonié. Many of the same species listed under *Diplothmema furcatum* were listed also under *P. furcata*.

The present specimen is placed under *Diplothmema furcatum* (Brongniart) Stur, since this species has priority to that of *Eremopteris missouriensis* Lx. and *Palmatopteris furcata* (Brongn.) Potonié.

Diplothmema obtusiloba (Brongniart) Stur
Plate 12, figure 3

1829. *Sphenopteris obtusiloba* Brongniart, Prodrome d'une histoire des végétaux fossiles, p. 204, pl. 53, fig. 2.

1833. *Sphenopteris irregularis* Sternberg, Versuch einer Geognostischen Botanischen Darstellung der Flora der Vorwelt, v. 2, fasc. 5–6, p. 63, pl. xvii, fig. 4; fasc. 7–8, p. 132.

1836. *Cheilanthites obtusilobus* (Brongn.) Goeppert, Systema, p. 246.

1836. *Cheilanthites irregularis* (Sternb.) Goeppert, Systema, p. 247.

1848. *Sphenopteris trifoliolata* (A r t i s ?) Brongn., SAUVEUR, Veg. foss. terr. houill. Belg., pl. xix, fig. 2; pl. xxi.
1869. *Sphenopteris (A n e i m i o i d e s) obtusiloba* (Brongn.) SCHIMPER, Traité de paléontologie végétale, v. 1, p. 399, pl. xxx, fig. 1.
1869. *Sphenopteris (Gymnogrammides) irregularis* (Sternb.) SCHIMPER, Traité de paléontologie végétale, v. 1, p. 373.
1877. *Diplothmema obtusilobum* (Brongn.) STUR, Die Culm-Flora, K.-k. geol. Reichsanstalt Abh., Band 1, Heft 2, p. 124 (230).
1877. *Diplothmema irregulare* (Sternb.) STUR, Die Culm-Flora, K.-k. geol. Reichsanstalt Abh., Band 8, Heft 2, p. 124 (230).
1884. *Pseudopecopteris obtusiloba* (Brongn.) LESQUEREUX, Coal Flora of Pennsylvania, v. 3, p. 753.
1893. *Sphenopteris (Pseudopecopteris) obtusiloba* (Brongn.) D. WHITE, U.S. Geol. Surv., Bull. 98, p. 52.
1938. *Sphenopteris whitii* BELL, Fossil Flora of Sydney Coalfield, Nova Scotia, p. 20, pl. I, figs. 2–5; pl. II, figs. 1–3.
1943. *Diplothmema o b t u s i l o b a* (Brongn.) Stur, WHITE, Lower Pennsylvanian species of *Mariopteris, Eremopteris, Diplothmema*, and *Aneimites* from the Appalachian Region, U.S. Geol. Surv. Prof. Paper 197-C, p. 97, pl. 30, fig. 4; pl. 35, figs. 7, 9.

Description.—Specimen at hand not well preserved, but satisfactory enough for determination. Pinnae opposite or nearly so, attached to a slightly striated rachis. Tertiary pinnae diverging from rachis at a nearly right angle, oblong-lanceolate. Pinnules alternate, cuneate, obtusely rounded, with short, broad stalks, or sessile and decurrent, bilobate or trilobate. Pinnules coriaceous, veins generally obscure.

Remarks.—The single specimen found during this study compares well with specimens figured by White (1943, p. 97, pl. 30, fig. 4; pl. 35, figs. 7, 9).

EREMOPTERIS BILOBATA White
Plate 12, figure 4

1899. *Eremopteris bilobata* WHITE, Fossil Flora Lower Coal Measures of Missouri, U.S. Geol. Surv., Mono. 37, p. 19, pl. IV, pl. V, figs. 4–6.

Description.—Pinnae alternate, triangular to oblong triangular, tapering from the base to the acute apex, distant. Pinnae closer to apex shorter and slightly wider than those farther down the rachis. Rachis bordered by the lamina from decurrent pinnules. Pinnules alternate, close to one another but not overlapping, decurrent, constricted at the base, ovate-deltoid when compound, truncate lobes. Basal pinnules bilobate. Lamina thin. Venation very poorly preserved in the materials at hand.

Remarks.—The single specimen found during the present study compares well with the drawings of enlarged pinnae shown by White (1899, pl. V, figs. 5a, 6a), but the photographs (pl. IV; pl. V, figs. 4, 5, 6) leave much to be desired and on the whole are completely worthless for diagnostic purposes. White mentioned in his description a blunt, spiny prolongation of the rachis similar to that of *Mariopteris muricata*. The prolongation is not definitely seen in the specimens at hand. The lobation is comparable to that of *E. cheathami* Lx. or of *Sphenopteris solida* Lesquereux. White questioned the establishment of *Eremopteris bilobata*. He felt that Lesquereux did not depict *S. solida* properly, since Lesquereux did not uncover the lobes of the pinnules or depict the rachial characters.

Jongmans and Dijkstra (1960, p. 1179) listed White's reference as the only occurrence of the species.

A perusal of White's 1943 publication did not show any other possible relationships. The present specimen is, therefore, cautiously referred to White's species, on the basis of pinnule form.

Sphenopteris mixta Schimper
Plate 13, figure 1

1869. *Sphenopteris (Cheilanthides) mixta* Schimper, Traité de paléontologie végétale, v. 1, p. 382.
1879. *Sphenopteris mixta* Schimper, Lesquereux, Coal Flora of Pennsylvania, Atlas, p. 9, pl. LIV, figs. 1–3a; text, v. 1 (1880), p. 276.
1884. *Pseudopecopteris nummularia* (Gutb.) Lesquereux, Coal Flora of Pennsylvania, v. 3, p. 751, pl. CIII, figs. 1–3; 2d type.

Description.—Rachis flexuous, striated longitudinally. Pinnules alternate, 3–7 mm long, 2–4 mm wide, triangular-linear, decurrent, distant to slightly overlapping, upper ones broadly attached, lower ones narrowly attached. Three to four ovate lobes on each side of the pinnule, close, separated by an acute sinus. Midrib flexuous, distinct, secondary branches departing nearly opposite each sinus; veins branching widely once or twice again before reaching margin.

Remarks.—White (1899, p. 35) retained Lesquereux's specimens of *S. mixta* (1879, pl. LIV, figs. 2 [1, 3?]) in that species, but qualified the acceptance of all the specimens figured. I agree with White's decision, as White had the opportunity to study the original specimens. The material at hand does not look like the specimens figured by Lesquereux, but compares well with White's figures (pl. XI, fig. 3, pl. XII, figs. 1–2). White found, in addition, that *Pseudopecopteris nummularia* (Gutb.) Lesquereux had no distinguishing specific characters from *S. mixta*. For this reason, *P. nummularia* was added to the synonymy list of *S. mixta* by White. After examining the description and figures shown by Lesquereux (1884, pl. CIII, figs. 1–3), I concur with White's decision.

Sphenopteris obtusiloba ? Brongniart
Plate 13, figure 2

1829. *Sphenopteris obtusiloba* BRONGNIART, Prodrome d'une histoire des végétaux fossiles, p. 204, pl. liii, fig. 2.

Description.—Fern pinna, linear-lanceolate; in those portions preserved, approximately 8.5 cm long, 2.5 cm wide, with obtusely rounded terminal pinnule. Pinnules alternate or opposite, with contracted, lobed, basal pinnules followed by bluntly rounded, slightly lobed to entire, decurrent pinnules to the apex of the pinna. Arched veins, very faint, appearing to diverge from a single vein entering pinnule at an acute angle from the striated rachis. Pinnules widest above midlength, except in those closest to terminal pinnule. In these pinnules, widest point is closest to the rachis, tapering to an obtuse point. Surface thick, coriaceous.

Remarks.—The specimens at hand compare fairly well with that shown by Janssen (1939, p. 111, fig. 90) and referred by him to *Sphenopteris obtusiloba* Brongniart.

When compared to material under the same epithet figured by Kidston (1923, pl. III) and Zeiller (1886, pl. III, figs. 1–4), there is a great deal of difference in the over-all size, the attachment of the pinnules, and the lobing. Kidston (1923, p. 30) pointed out the following:

> *As in all fronds which attain to large dimensions . . . the pinnules of* Sphenopteris obtusiloba *form no exception to the general rule of great variation in their size and extent of the pinnule segmentation, according to the position of the pinnule on the pinna, and also the position of the pinna that bears them on the frond. Hence considerable study is necessary to enable one to become familiar with the various forms of pinnules that occur on different portions of the frond of* Sphenopteris obtusiloba.

The specimens in the present study are tentatively assigned to *Sphenopteris obtusiloba* on the basis of the general characters and because of the variation that seems to characterize the species.

SPHENOPTERIS sp. Brongniart

Plate 13, figure 3

1822. *Filicites* (sect. *Sphenopteris)* BRONGNIART, Class. de Végét. Mem. mus. hist. nat., v. 8, p. 233.
1828. *Sphenopteris* BRONGNIART, Prodrome d'une histoire des végétaux fossiles, p. 28.

Description.—Fern frond portion, rachis striated, thin, flexuous. Primary pinna triangular-lanceolate, tapering to an acute point. Secondary pinnae turned upward, alternate, oblong-lanceolate, bluntly pointed, distant to somewhat overlapping. Pinnules alternate, bluntly rounded, longer than wide, oblique to the rachis, decurrent, no more than 2 mm long, approximately 1 mm wide in portions preserved. No venation observable.

Remarks.—These specimens were found only in a dark gray shale in the Clary Pit, Deepwater, Missouri. The material was badly preserved. No venation was observable; even the general outlines are macerated to a degree. Therefore, the material at hand is not referred to any species. The closest comparison is with the specimen figured by White (1899, pl. XIX, figs. 2, 2a) and referred by him to *Sphenopteris* cf. *gravenhorstii* Brongniart. As White pointed out, this was a very questionable assignment. There appears to be some conflict as to the separation of the entities involved in the *S. gravenhorstii–S. chaerophylloides–S. cristata* complex. White (1899, p. 50) considered a thorough revision of this group of Sphenopterids necessary in order to distinguish which species had been identified from the vicinity of Clinton, Missouri.

ASTEROTHECA Presl sp.?
Plate 14, figure 1

Description.—Pinnules at least three times longer than broad, alternate. Fructifications borne on each side of the midrib along the pinnule margins. Venation poorly preserved on the specimens at hand. Fructifications oblong, clustered, apparently united at the base to form small groups of four to five sporangia. No pedicel apparent.

Remarks.—The fertile material found during the

course of this study was very poorly preserved. Even applications of xylene did not permit species determination. Slides of peels revealed structures similar to those figured by Gothan and Remy (1957, p. 141) for *Asterotheca* Presl.

White (1899, p. 89) described some sporangia that he found on Missouri samples, and the descriptions compare very favorably with the present material. White went on to say that he sent samples of his fertile material to Professor Zeiller, who considered White's specimens as somewhat intermediate between *Asterotheca* and *Scolecopteris*, though probably much nearer *Asterotheca*.

Mariopteris (Pseudopecopteris) decipiens (Lesquereux) White
Plate 14, figure 2; Plate 15, figure 1

1854. *Sphenopteris decipiens* LESQUEREUX, Boston Jour. Nat. Hist., v. 6, p. 420.

1879. *Pseudopecoteris decipiens* LESQUEREUX, Coal Flora of Pennsylvania, Atlas, pl. LII, figs. 9–10a, text (1880), p. 214.

1893. *Mariopteris (Pseudopecopteris) decipiens* (Lesquereux) WHITE, U.S. Geol. Surv., Bull. 98, p. 47, pl. I, figs. 5–8, 5a; pl. II, figs. 1–3, 3a.

Description.—Pinnae subopposite, at right angles to the rachis or oblique, touching or slightly separated, in those parts preserved 3.3 cm long and 1.1 cm wide. Pinnules decurrent, connate by a narrow, decurring lamina, obovate, with rounded tips alternate, slightly overlapping, 5–7 mm long, 3–6 mm wide, the lowest pinnules broader, sometimes orbicular at the base of the pinna, some bilobate with two unequal lobes.

Venation generally not visible except in the lower part of the pinnule. In some of the lower pinnules the primary vein rather prominent, in others, the primary vein not distinct; primary vein departing from the rachis at a very acute angle and arching and branching gradually.

Remarks.—The specimens at hand agree very well with the material figured by Lesquereux (1879, p. 214, pl. LII, figs. 9–10a) and by White (1893, p. 47, pl. I, figs. 5–8, 5a; pl. II, figs. 1–3, 3a).

White placed the specimens in the genus *Mariopteris*, rather than *Pseudopecopteris* as Lesquereux had done, because he felt the broad, flat rachises, which dichotomized below, indicated their position in *Mariopteris*. White (1893, p. 49) was quite hesitant about referring his specimens to *Mariopteris (Pseudopecopteris) decipiens*. He concluded that the species, on the basis of descriptions and figures, was either extremely polymorphous or quite ill-defined. He felt also that his Missouri materials would represent a new species, if it should be necessary to separate it from *Pseudopecopteris decipiens* because of its slight agreement with figures and descriptions of the type material. This revision and redefinition of the species would necessitate examination of the types and the material originally studied by both Lesquereux and White and definitely should be done at some later date. Until such a re-examination can be made, I am of the opinion that the designation *Mariopteris (Pseudopecopteris) decipiens* (Lx.) D. White is justified.

MARIOPTERIS SPECIOSA (Lesquereux) White

Plate 15, figure 2; Plate 16, figure 1

1879. *Pseudopecopteris speciosa* LESQUEREUX, Coal

Flora of Pennsylvania, Atlas, pl. LI, figs. 1-1b, text (1880), p. 216.
1893. *Mariopteris speciosa* (Lesquereux) WHITE, U.S. Geol. Surv., Bull. 98, pp. 47, 49, 52.

Description.—Pinna 4.5 cm long, 1.5 cm wide at its widest point, tapering down to its apex gently, lanceolate. Pinnules separate or slightly overlapping, alternate; upper pinnules obliquely attached, almost horizontal in those lower; ovate or ovate-lanceolate, obtusely pointed; lower pinnules three- to five-lobed, attached by a small portion of their bases, the upper ones decurrent. Midrib prominent, decurrent from the rachis.

Remarks.—There appears to be a great deal of variability in this species, as shown by the material figured by Lesquereux (1879, pl. LI, figs. 1-1b) and the specimens from the present study.

Lesquereux (1880, p. 216) pointed out that there was a close comparison of *Mariopteris speciosa* to *Pseudopecopteris latifolia*, *P. acuta,* and *P. nervosa*. Jongmans and Dijkstra (1960, p. 1591) suggested a possible comparison to *M. soubeirani*. In regard to the latter, although Kidston (1925, p. 649) placed *M. speciosa* in synonymy with *M. soubeirani* Zeiller, Kidston inclined to the view that *M. soubeirani* was specifically distinct from *M. speciosa* Lesquereux, but that it was impossible to determine the relationship of the two species without a comparison of American examples with European specimens. Kidston stated that Zeiller saw the close relationship between the two species but felt that the tertiary pinnae were more triangular in *M. soubeirani*.

Comparison of the pinnae of *M. speciosa* with *P.*

latifolia shows that the pinnae are quite distantly separated in the latter, although the general configuration of the pinnules is the same. The apex of the pinna in *P. acuta* is pointed and differs significantly enough from *M. speciosa* to make them readily separable. *P. nervosa* possesses long, oblong, obtuse pinnules comparable to some *Alethopteris* species.

It is very difficult to make identifications or comparisons based on Lesquereux's terse descriptions and illustrations, and therefore no effort is made to relate the present specimens to species other than the one to which it is referred.

MARIOPTERIS SPHENOPTEROIDES (Lx.) Zeiller
Plate 16, figure 2

1879. *Odontopteris sphenopteroides* LESQUEREUX, Coal Flora of Pennsylvania, Atlas, p. 4, pl. XXI, figs. 3, 4; text, v. 1 (1880), p. 139.

1886. *Mariopteris sphenopteroides* (Lx.) ZEILLER, Fl. foss. houill. Valenciennes, Atlas, pl. XIX, figs. 3, 4; text (1888), p. 171.

Description.—Pinnae a l t e r n a t e , close together, oblong-triangular, tapering to an acute point; rachis striated. Pinnules alternate, ovate-deltoid, divided into lobes or teeth; upper pinnules decurrent, with well-formed sinuses between; lobes or teeth turned upward. Single midrib from the rachis, forking and supplying each lobe with at least two ultimate veins.

Remarks.—The present specimens, although rather fragmentary, compare well with White's descriptions and figures (1899, pp. 31–33, pl. IX, figs. 1, 2; pl. X). The material also corresponds with Lesquereux's and Zeiller's figures. Of the former, the illustrated venation pattern leaves much to be desired, but White pointed out that this was due entirely to imperfect drawing. White noted that the species had been found in the western coal region of Arkansas and in the Lower Productive Coal Measures of Mazon Creek, Illinois. This characterizes a limited zone near the base of the Allegheny Series.

White (1899, p. 31) united *M. nobilis* Achepohl with *M. sphenopteroides* Lesquereux. Kidston (1925, p. 640) commented that the two plants had a great similarity but that the ultimate pinnae of White's plant were much longer and more linear.

PECOPTERIS CLINTONI Lesquereux

Plate 16, figure 3; Plate 17, figure 1

1879. *Pecopteris clintoni* LESQUEREUX, Coal Flora of Pennsylvania, Atlas, p. 8, pl. XLII, figs. 1–4a (not figs. 5, 5a, 5b); text, v. 1 (1880), p. 251 (pars).

Description.—Pinnae linear-lanceolate, s l i g h t l y overlapping at their margins, terminated by an obtusely pointed terminal pinnule that is united with the adjacent pinnules. Pinnules alternate, lateral margins slightly separate, decurrent, connate, oblong, with obtuse ends; very thin lamina. Midrib decurrent, becoming indistinct at about three quarters the length of the pinnule; lateral veins fine, dividing near the base, dividing again and meeting the margin obliquely.

Remarks.—The material at hand, although small in amount, compares well with specimens figured by White (1899, pl. XXXIV, pl. XXXV, fig. 4). The villosity mentioned by White is not visible in the specimens collected for this study.

White (1899, pp. 94, 95) noted that Lesquereux erroneously referred certain specimens of *P. clintoni* to *Callipteridium membranaceum*; the specimens either lacked the alethopterid nervation or were not properly delineated by the artist, in White's opinion.

PECOPTERIS DENTATA Brongniart
Plate 17, figure 2; Plate 18, figure 1

1828. *Pecopteris dentata* BRONGNIART, Prodrome d'une histoire des végétaux fossiles, p. 58, 170.
1828. *Pecopteris plumosa* (Artis?) BRONGNIART, Prodrome d'une histoire des végétaux fossiles, p. 58, 171.
1879. *Pecopteris pennaeformis* B r o n g n i a r t, LESQUEREUX, Coal Flora of Pennsylvania, Atlas, p. 8, pl. XLV, figs. 1, 1a (figs. 2, 2a?); text, v. 1 (1880), p. 239.
1883. *Dactylotheca dentata* (Brongn.) ZEILLER, Ann. Sci. Nat. (6) bot., v. 16, p. 184, pl. ix, figs. 12–15.

Description.—Pinnae at an oblique angle to the rachis, overlapping, 4.0 cm apart at the rachis, wider at middle than at base. Secondary pinnae alternate to subopposite, approximately 6 mm apart, upper ones oblique; middle pinnae nearly at right angles; lower pinnae sometimes reflexed and shorter, linear-lanceolate, 1.2–3.5 cm long, 3–7 mm wide, tapering gently to an obtusely acuminate point. Pinnules alternate, more or less triangular, somewhat arched, obtusely pointed.

Remarks.—When compared with Lesquereux's (1879, p. 240, pl. XLIV, figs. 4, 4a) figures of *P. dentata*, there is very poor agreement, but his figures of *P. pennaeformis* (pl. XLV, figs. 1–2a) are in complete agreement with the present material. White (1899, p. 76) said Lesquereux's material was *P. dentata*. There is excellent agreement with the specimens figured by White (1899, pl. XXIV, figs. 1–2; pl. XXV; pl. XXVII). The general area of collection, Henry County, Missouri, was the area from which White's material was obtained.

The material at hand also compares well with the specimens figured by Zeiller (1886, pl. XXVI, figs. 1–2E; pl. XXVII, figs. 1–4; pl. XXVIII, figs. 5, 5a) and referred by him to *Pecopteris (Dactylotheca) dentata* Brongniart.

Jongmans and Dijkstra (1962, p. 2260–2265) provided an excellent review of the species, a large number of which were taken from White's rather extensive list, assigned to comparable specimens. There appears to be a gradation of characteristics ranging from *P. plumosa* to *P. dentata*; many authors refer to the complex as *P. plumosa-dentata* or *P. dentata-plumosa*.

Arnold (1949, p. 187) mentioned that Kidston merged *P. dentata* with *P. plumosa,* because on the large and well-preserved fronds from the British Coal Measures both *dentata* and *plumosa* types are present as variants of different parts of the same frond.

PECOPTERIS PSEUDOVESTITA White
Plate 17, figure 3; Plate 18, figure 2

1879. *Alethopteris ambigua* LESQUEREUX (in part), Coal Flora of Pennsylvania, Atlas, p. 6, pl. XXXI, figs. 2, 3 (4?); text, v. 1 (1880), p. 182.
1879. *Pecopteris clintoni* LESQUEREUX (in part), Coal Flora of Pennsylvania, Atlas, p. 8, pl. XLII, figs. 5, 5a–b; text, v. 1 (1880), p. 251.
1879. *Pecopteris vestita* LESQUEREUX (in part), Coal Flora of Pennsylvania, p. 8, pl. XLIII, figs. 5, 5a?; text (1880), p. 252.

1899. *Pecopteris pseudovestita* WHITE, Fossil Flora of the Lower Coal Measures of Missouri, U.S. Geol. Surv., Mono. 37, p. 85–91, pl. XXVIII, figs. 1–2a; pl. XXIX, pl. XXX, pl. XXXI, figs. 1, 2, 3?; pl. XXXII, figs. 1, 2.

Description.—Pinnae a l t e r n a t e , long, linear-lanceolate, with an obtuse apex, almost at right angles to the rachis, sides subparallel, tending to curve upward slightly, close or slightly overlapping; terminal pinnule ovate. Pinnules alternate, slightly oblique to the rachis, small, 3.5–5 mm long, 2 mm wide, joined to adjacent pinnules for about one half their length, sides almost parallel, apex rounded; lowest pinnules slightly reduced, the uppermost pinnules joined to the terminal pinnule. Midrib distinct, slightly decurrent at base, lateral veins not well defined in the specimens at hand.

Remarks.—White (1899, p. 85–91) united under *P. pseudovestita* certain specimens of *P. clintoni, Alethopteris a m b i g u a,* and *Callipteridium membranaceum* that Lesquereux had identified. White felt that not only had the same forms been referred to two different species, but forms belonging to more than one species had been included under each name.

There is good agreement between the present specimens and those shown by White (1899, pl. XXVIII, figs. 1, 2; pl. XXIX, pl. XXXII, fig. 1).

CYCLOPTERIS TRICHOMANOIDES Brongniart
Plate 19, figures 1 & 2

1830. *Cyclopteris trichomanoides* BRONGNIART, Prodrome d'une histoire des végétaux fossiles, I. 5, p. 217, pl. 61, fig. 4.

1833. *Cyclopteris dilatata* LINDLEY AND HUTTON [non (L. & H.) Sternb.], Fossil Flora, v. 2, pl. xciB.
1880. *Neuropteris dilatata* (L. & H.) LESQUEREUX, Coal Flora of Pennsylvania, v. 1, p. 78.
1893. *Neuropteris dilatata* (L. & H.) LESQUEREUX, D. White, Bull. U.S. Geol. Surv., No. 98, p. 96.

Description.—Reniform-orbicular pinnules, large, greatest diameter 2.9–3.7 cm; auriculate, usually a smaller lobe on one side than on the other, attachment point off center. Venation variable, ranging from coarse to fine, radiating from the point of attachment, dichotomizing two to four times before reaching the margin at almost a right angle when opposite the point of attachment, but forming acute angles with the margin along the lateral edges of the lamina.

Remarks.—There is a great deal of variation found in *Cyclopteris* pinnules. Many of these have been found in direct association with *Neuropteris* pinnules and have been assigned to them as rachial pinnules. Janssen (1939, p. 157) stated:

> Nearly every leaf is slightly different from the next, and various species have been made by authors to accommodate such differences. Such species names are of little value, and it is more satisfactory to refer all detached specimens to C. trichomanoides *which has priority as a form name.*

Because of the generally accepted opinion that Cyclopterid pinnules have no stratigraphic value, Janssen's suggestion bears merit in relieving the problem of such a mass of names.

White (1893, p. 98) pointed out that American species of *N. dilatata* appeared to conform more closely to the description and characters of *C. trichomanoides*. His 1899 publication showed figures (pl. XLI, fig. 6; pl. XLII, figs. 1, 1a; pl. XLIII; pl. XLIV, fig. 2) of *N. dilatata*, which, in my opinion, should be included under *C. trichomanoides*.

CYCLOPTERIS sp. Brongniart
Plate 24, figure 1

Description.—Unattached cyclopterid pinnules; variable in size, but generally small, oval, oblong, or round. Leaf margin entire. Attachment point close to center. Venation radiating from point of attachment.

Remarks.—No effort is made here to relate these unattached cyclopterid pinnules to a specific entity. There is no way to determine whether they are basal pinnules or rachial pinnules.

LINOPTERIS Presl

1835. *Dictyopteris* GUTBIER. Abdrücke und Versteinerungen des Zwickauer Schwarzkohlengerbirges, p. 62.

1838. *Linopteris* PRESL, in Sternberg, Versuch einer Geognostischen Botanischen Darstellung der Flora der Vorwelt, v. 2, fasc. 7–8, p. 167.

LINOPTERIS GILKERSONENSIS White
Plate 19, figure 3

1897. *Dictyopteris* sp. WHITE, Geol. Soc. Amer. Bull., v. 8, p. 297, 301.

1899. *Dictyopteris gilkersonensis* WHITE, 19th Ann. Rept. U.S. Geol. Surv., pt. 3, p. 510.

1899. *Linopteris gilkersonensis* WHITE, Fossil Flora of the Lower Coal Measures of Missouri, U.S. Geol. Surv., Mono. 37, p. 139, pl. 41, figs. 7, 8.

Description.—Detached pinnules 14 mm in length, 5–5.5 mm in width, lateral margins nearly straight and parallel, leading to a rounded apex. Base nearly equilateral. Midrib moderately prominent, anastomosing secondary veins very thick and raised. Areoles few.

Remarks.—The specimens at hand compare favorably with those figured by White (1899, p. 139, pl. XLI, figs. 7, 8; pl. LXI, fig. 1f) also from the Gilkerson's Ford collection area. White (1899, p. 204) differentiated *L. gilkersonensis* from *L. munsteri* Kidston and Zeiller on the basis of "the stronger nervation, rigid and open angled nerves, with shorter meshes at the margins." Furthermore, he separated it from *L. obliqua* Bunbury because of its "straighter pinnules, the coarse veins, and the relatively few and broad meshes." Crookall (1959, p. 204, 208) accepted White's separation and his erection of the new species as valid and justified.

NEUROPTERIS (Brgt.) Sternberg

1822. *Filicites* (Sect. *Nevropteris*) BRONGNIART, Class. de Végét. Foss. Mem. Mus. d'hist. nat., v. 8, p. 233.

1825. *Neuropteris* STERNBERG, Versuch einer Geognostischen Botanischen Darstellung der Flora der Vorwelt, v. 1, fasc. 4, p. xvi.

NEUROPTERIS CAUDATA White
Plate 19, figure 4

1893. *Neuropteris caudata* WHITE, U.S. Geol. Surv., Bull. 98, p. 87-91, pl. IV, figs. 1-9.

Description.—Rachis striate. Pinnules coriaceous, alternate, attached to the rachis at a right angle, the upper pinnules somewhat oblique, contiguous below, becoming more distant above, 9–10 mm long, 4–5 mm wide, rounded at the base above the point of attachment, the lower angle prolonged downward in an acuminate, spurlike appendage. Pinnule margins parallel in the middle, upper border straight or slightly concave, lower border curving up to meet upper border at a more or less upturned tip, which may be obtuse or rounded. Upper pinnules attached by a greater distance of their base, becoming decurrent near the pinna apex.

Remarks.—The specimens at hand are referred to White's species on the basis of the awnlike prolongation of the lower auricle, which appears to be a rather prominent and constant characteristic. The venation of the present specimens is poor, but as White (1893, p. 88) pointed out, the nervation is nephropteroid—the midrib consists of a band of vascular bundles forking and diverging, the secondary veins obscure in the thick textured pinnules. Such appears to be the case with the specimens at hand.

NEUROPTERIS ELACERATA sp. nov.

Plate 20, figure 1

Description.—Oblong neuropterid pinnules, amount preserved 2.9–3.1 cm long, 0.8–1.3 cm wide, exhibiting fimbriae 0.1–1.1 cm in length. Fimbriae usually tapering from the base to a sharp point, each supplied with one vascular bundle, the fimbriae appearing first above the middle of the pinnule. Margins of the pinnule parallel or almost so, up to the point of fimbriation. Upper angle (lobe), formed between each fimbra and the main axis, rounded. Fairly prominent groove in midrib region running for a greater length of the pinnule; lateral veins departing from midrib at an angle of 45° or less, arching to the margins. Lateral veins, numbering from 18–22 per cm along the margin, where they meet it at about 45°. Lower veins departing at a smaller angle than the more distal ones.

Remarks.—The first impression gained from the single specimen at hand is that these are neuropterid pinnules that have undergone a great deal of rough treatment. These pinnules do not compare with Lesquereux's specimens, as shown by Crookall (1959, p. 151, fig. 49), which were assigned to *Cyclopteris fimbriata*. They were first figured in 1858.

The present specimens are not cyclopterid, nor are they ovate in shape. Crookall (1959, p. 150) stated fimbriated cyclopterids probably all belonged to *Neuropteris ovata* Hoffman *sensu generali*.

Because of the difference in shape of the pinnules and fimbriae of the present material from those figured by many other investigators, such as Lesquereux (1858, 1880), White (1893), Bertrand (1930), and Crookall (1959), it seems advisable to describe the Missouri specimen as new. The possibility that the condition of the pinnules was caused by poor preservation is ruled out because the lobes of the lamina between fimbriae are rounded and not torn or lacerated.

These specimens were not described in White's Flora probably because any specimens of this sort, if they were found, were not considered as perfect specimens and may have been thrown away.

Type specimen.—No. 913
Locality.—Clary Pit, Deepwater, Missouri
Repository.—Paleobotanical Collection, University of Missouri

Neuropteris heterophylla Brongniart

Plate 20, figure 2; Plate 21, figure 1

1822. *Filicites (Nevropteris) heterophyllus* Brongniart, Class. de Végét. Foss., Mem. mus. hist. nat., p. 33, 239, pl. ii, figs. 6a, 6b.

1828. *Neuropteris heterophylla* Brongniart, Prodrome d'une histoire des végétaux fossiles, p. 53.

Description.—Secondary pinnae, opposite, oblique; margins parallel for a considerable distance; up to 5.5 cm long by 1.5 cm broad; distant or slightly touching by lateral margins; linear-lanceolate; terminal pinnule elongated, slender, with gradually tapering margins and obtusely pointed apex. Rachis wide and striated. Pinnules alternate, elongate-triangular, often slightly falcate; the two lateral pinnules situated near the terminal pinnule attached by a broad base, most lateral pinnules attached by a single point, sessile, more or less cordate at the base with lateral margins converging slightly towards the rounded apex; slightly separated. Midrib strong, extending up close to apex; lateral veins thin but distinct, arising at a sharp angle, arched to the margin.

Remarks.—There is a great deal of variability in this species. The following is a partial list of species that have at one time or another been referred to *Neuropteris heterophylla,* but I have been unable to form an opinion concerning their position:

N. *loshii* Brongniart

N. *microphylla* Brongniart

N. *polymorpha* Dawson

N. *thymifolia* Sternberg

N. *eltringhami* Kidston

N. *camptophylla* Jongmans and Gothan

N. heterophylla probably is most easily confused with *N. tenuifolia* Schlotheim, but the lateral veins of the former are stronger and fewer than are those of the latter. *N. heterophylla* may be distinguished from *N. rarinervis* Bunbury by the finer and less prominent, but more numerous and more divided, lateral veins.

Crookall (1959, p. 104) stated that *N. heterophylla* is one of the commonest and most widespread Coal Measure plants. It is fairly common in the Westphalian A; very common in the Westphalian B; fairly rare in the Westphalian C; and very rare in the Westphalian D.

NEUROPTERIS MISSOURIENSIS Lesquereux
Plate 22, figure 1

1879. *Neuropteris missouriensis* LESQUEREUX, Coal Flora of Pennsylvania, Atlas, p. 3, pl. vii, figs. 5–6, 6a; text, v. 1 (1880), p. 104.

Description.—Pinnae alternate, almost at right angles or slightly oblique to the rachis, which is thick. Lateral pinnules alternate, slightly oblique to the pinnae rachis, oval, obtusely rounded at the apex, overlapping each other. Midrib distinct, preserving identity to at least the midpoint of the pinnule. Lateral veins arising from the midrib at a sharp angle, forking near the base, arching to meet the margin at about 75°–80°.

Remarks.—White (1899, p. 131) felt that *N. missouriensis,* which is very similar to *N. flexuosa* Stb., could be distinguished from the latter by its open pinnae, the oblong or oval, slightly imbricated pinnules, and the distinct nervation.

The present material agrees well with that figured

by White (1899, pl. XLI, fig. 4; pl. XLII, fig. 4) and by Lesquereux (1879, pl. VII, fig. 5).

NEUROPTERIS OVATA Hoffmann forma FLEXUOSA (Sternberg) Crookall
Plate 23, figures 1 & 2

1826. *Neuropteris flexuosa* STERNBERG, Essai d'un exposé géognostico-botanique de la flore du monde primitif, v. 1, fasc. 3, p. 44, pl. XXXII, fig. 2.
1879. *Neuropteris vermicularis* LESQUEREUX, Coal Flora of Pennsylvania, pl. X, figs. 5–10; text, v. 1 (1880), p. 99.
1959. *Neuropteris ovata* forma *flexuosa* (Sternberg) CROOKALL, Foss. pl. of Carb. Rocks Gr. Brit., 2d Sect. Palaeontol., v. 4, no. 2, p. 158, pl. XXVII, figs. 1–3; pl. XXXVIII, figs. 1–3; pl. XLIX, figs. 5, 6; pl. L, figs. 3–5; text, figs. 52, 62J, 66C, J.

Description.—Lateral pinnules 1.3–2.1 cm long, 0.5–0.8 cm wide, well-rounded apex, upper basal angle rounded, lower basal angle typically auriculate, linear-lanceolate, suboppositely arranged, attached to the rachis at an angle of 45°, slightly overlapping or slightly separated; terminal pinnule 3 cm long, 1.2 cm wide. Midrib of pinnules prominent for over one half the length, lateral veins arising at a sharp angle, arching and dividing four times. Pinna rachis thin and slightly angled between pinnule attachment points.

Remarks.—The specimen at hand compares well with that figured by Wood (1963, p. 112, pl. 9, fig. 6) and White (1893, pp. 91–93, pl. V, figs. 1–5) and assigned by them to *N. flexuosa* Sternberg.

The present specimen also compares with Crookall's (1959, p. 158–162, pl. XXXVII, fig. 1 and text fig. 66C, J) figures except in the shape and length of the terminal pinnule. In general, the present specimens are larger.

As pointed out by other authors, there is a great deal of variability in this group, but the pinnule shape and venation appear to be rather constant. Because of the variability allowed for within Crookall's assignment, the present material is placed in the species and form.

Crookall (1959, p. 155) claimed that *N. flexuosa* Sternberg corresponded to *N. vermicularis* Lesquereux. Lesquereux's plate (1879, pl. X, figs. 5–10) shows a terminal pinnule, but the material at hand does not compare closely with that illustration. However, the pinnules shown by Lesquereux (1879, pl. X, fig. 6) do compare well with present samples. Dijkstra and Jongmans (1961, p. 1851) noted the comparison between *N. vermicularis* Lx. and *N. ovata* forma *flexuosa*.

The present specimens, in regard to rachis and terminal pinnule, compare with the specimens figured by Lesquereux (1879, pl. XII, figs. 1, 3) and assigned to *N. tenuifolia* Lx. Bolton (1926, p. 307) found a series of intermediate forms linking *N. tenuifolia* Schloth. (sp.), *N. flexuosa* Sternb., and *N. gigantea* Brongniart. Bolton found the uppermost pinnules of a specimen that undoubtedly belonged to *N. tenuifolia*; the lower ones had the characters associated with *N. flexuosa*.

Crookall (1959, p. 120) noted that pinnules bearing *N. flexuosa* and *N. tenuifolia* characteristics had been found in Missouri and that these variations very strongly indicated a specific unity of the American species in a single species that is possibly distinct from European species. The present material would lend support to this theory.

NEUROPTERIS SCHEUCHZERI Hoffmann
Plate 24, figure 1

1691. *Phyllites mineralis* LUIDIUS, Lithophyl. Brit., p. 12, pl. v, fig. 190.
1723. *Phyllites mineralis* Luid., SCHEUCHZER, Herb. Dil., p. 48, pl. x, fig. 3.
1826. *Neuropteris scheuchzeri* HOFFMANN, in Keferstein, Teutschland geogn.-geol. Dargest., v. 4, p. 157, pl. 1b, figs. 1–4.
1830. *Neuropteris angustifolia* BRONGNIART, Prodrome d'une histoire des végétaux fossiles, p. 231, pl. lxiv, figs. 3, 4.
1832. *Neuropteris cordata* (Brongniart), LINDLEY AND HUTTON, Fossil Flora, v. 1, p. 119, pl. xli.
1854. *Neuropteris hirsuta* LESQUEREUX, Boston Jour. Nat. Hist., v. vi, 4, p. 417.

Description.—Pinnules large, 5.8–7.5 cm long, lanceolate, sides straight or tapering unequally, obtusely tipped to sharply pointed, base lobed. Midrib variable, prominent the entire length of the pinnule or, as in the figure shown, prominent only one fourth the length, veinlets numerous, branching from the midrib and arching toward the margins, forking one to five times. Numerous short hairs present on the pinnule surface.

Remarks.—All of the specimens in this study assigned to this species are isolated pinnules. As can be seen by the above synonymy list, a variety of species of neuropterid leaves have been grouped together, giving wide variation to those pinnules within the group. Wood (1963, p. 60), in his remarks, discussed but excluded the addition of *N. fasciculata* Lx., which Janssen (1940, p. 45, 46) concluded did not constitute a distinct species but should be referred to *N. scheuchzeri*. Wood also noted that Janssen added *N. decipiens* to the list. In referring to Janssen's study (1940, p. 46), the impression is gained that Sellards (1908, p. 414) divided Lesquereux's Kansas specimens of *N. fasciculata* into *N. scheuchzeri* and *N. decipiens*. Janssen went on to state that "the distinction between the two latter species is not clear cut, and there is a tendency among some authors to unite them." Jongmans (1961, p. 1823) felt that perhaps *N. decipiens* belonged under *N. scheuchzeri* as a large form of the latter. Specimens of *N. decipiens* were not encountered in this study.

Gothan and Remy (1957, p. 135) placed *N. scheuchzeri* Hoffm. under the genus *Paripteris,* which they erected to encompass all Neuropterids that were terminated by two pinnules and that did not belong to the Medullosan ferns. I believe that, since a valid genus has been recognized for so many years, the erection of a new genus, based on the pinnule arrangement on a pinna, is not as useful as it might appear, since much of our material consists of isolated pinnules whose arrangement on the pinna is, in most cases, unknown.

HOLCOSPERMUM Nathorst sp.?
Plate 19, figure 5

1914. *Holcospermum* NATHORST, Nacht. zur Palaoz. Flora Spitzbergens.

Description.—Seed, 2.5 cm long, 1.7 cm wide, flat.

Remarks.—The greater part of the organic material has been lost from this compression, making the identification of this single specimen difficult. The specimen at hand is tentatively referred to Nathorst's genus, on the basis of its general shape and size.

Janssen (1939, p. 164) pointed out that seeds of this type have been found in close association with pinnules of *Neuropteris scheuchzeri* and that possibly one form of *Holcospermum* may have been borne by this seed fern. The single specimen found in this study was embedded with *N. scheuchzeri,* as well as with other fern types, so that Janssen's statement is neither disproved nor proved by this study.

There is fair agreement between the present specimen and that shown by Wood (1963, pl. 11, fig. 1) and by Canright (1959, pl. 5, fig. 5), except that the striations are lacking or not preserved in the present specimen.

Spiropteris Schimper sp.?
Plate 21, figure 2

Remarks.—The specimen shown was the only *Spiropteris* sp. found. The specimen is placed in the form-genus *Spiropteris* rather than within one of the other fern genera based on leaf morphology because it is impossible to determine the details of the pinna, due to the circinate vernation.

Artisia transversa (Artis) Sternberg
Plate 9, figure 1

1825. *Sternbergia transversa* Artis, Antediluvian Phytology, 24p., 24 pls., London.

1838. *Artisia transversa* (Artis) STERNBERG, Essai d'un exposé géognostico-botanique de la flore du monde primitif, v. 2, pts. 7–8, p. 192.

Description.—Greatly flattened pith cast showing a number of horizontally aligned grooves, which are not parallel. Horizontal segments approximately 16 mm wide, height varying from 1.5–3.0 mm. Bordering strips on each side of the horizontal segments approximately 1 mm thick.

Remarks.—The discoid type of pith cast has been referred to Cordaitean stems. They were interpreted by Janssen (1939, p. 73) as rupture lines in the pith resulting from failure of the pith to elongate as rapidly as the rest of the stem. The material at hand compares favorably with that figured by Janssen (1939, fig. 56). This specimen is not so deeply marked as the one shown by Wood (1963, pl. 11, fig. 8) or by Canright (1959, pl. 5, fig. 10) and might be interpreted as a less mature specimen or one that had not undergone as much degradation prior to burial.

CARDIOCARPUS CRASSUS Lesquereux
Plate 9, figures 3 & 4

1884. *Cardiocarpus crassus* LESQUEREUX, Coal Flora of Pennsylvania, v. 3, p. 812, pl. CIX, fig. 12, pl. CX, figs. 6–9.

Description.—Seed, small, broadly oval, 4–4.5 mm long, 3–3.5 mm wide; wing approximately 0.5 mm wide; surface smooth, except for some folds running the length of the nucellar region. Wing narrowing at micropylar opening.

Remarks.—The specimens at hand are referred to *C. crassus* on the basis of the general agreement to

Lesquereux's (1884, p. 812, pl. CIX, fig. 12, pl. CX, figs. 6–9) description and figures. I agree with Lesquereux's statement concerning the resemblance of *C. crassus* to *C. simplex*. There is a certain amount of confusion associated with Lesquereux's citations. Jongmans and Dijkstra (1958, p. 507) pointed out that Lesquereux, in his Atlas, referred pl. LXXXV, figs. 48, 50 to *C. simplex*. In Lesquereux's text, p. 509, *C. simplex* is cited as being pl. LXXXV, figs. 49, 50. On the other hand, *C. diminutivus,* p. 570, is cited as being found on pl. LXXXV, fig. 48.

As pointed out by Lesquereux (1884, p. 812), it is difficult to separate species of *"Samaropsis."* His judgment that *C. zonulatus, C. late-alatus,* and *C. crassus* may represent one species is very plausible, since *C. late-alatus* has been identified in this flora. The specimens identified as *C. crassus* here may represent merely a less mature stage of growth of the former. For the present, the two species are treated here as separate entities until some further investigation shows them to have been attached to the same plant. They should then be united.

CARDIOCARPUS LATE-ALATUM Lesquereux
Plate 9, figure 2

1879. *Cardiocarpon late-alatum* LESQUEREUX, Coal Flora of Pennsylvania, p. 568, Atlas, pl. LXXXV, figs. 46, 47.

Description.—Flat seed, round to ovate, 7 mm wide by 9 mm long. Almost circular nucule without indentation at base, but obtusely pointed at apex. Apex of seed wing more acutely pointed than nucule, due to broadening of wing along micropylar sinus. Wing approximately 1 mm wide along the chalazal and lateral portions, but thickening to 2 mm in apical region.

Remarks.—These specimens, found exclusively at Gilkerson's Ford, compare favorably with that figured by Lesquereux (1879, Atlas, pl. LXXXV, figs. 46, 47) for *Cardiocarpon late-alatum*. The specimen shown by Arnold (1949, p. 229, pl. XXXIV, fig. 2) is only slightly larger, but looks identical.

Cardiocarpon was originally described by Brongniart in 1828. In 1864-65, Goeppert altered the spelling of Brongniart's *Cardiocarpon* to *Cardiocarpus*. Jongmans (1958, p. 489) referred the two genera to *Cardiocarpon (Cardiocarpus)* Brongniart. *Cardiocarpus* Brongniart is now the conserved name.

It is interesting to note that White (1899, p. 266) also found abundant seeds at Gilkerson's Ford, the same site from which this specimen was recovered, but referred them to *Cardiocarpon branneri* Fairch. and D. White. There is considerable agreement between the specimens at hand and those figured by Lesquereux, White, and Arnold. Further study is strongly indicated to determine whether there is enough variability in the two species to suggest their retention as separate entities. If such is not the case, then the epithet *C. late-alatum* should be retained, because of priority.

CARDIOCARPUS OVALIS Lesquereux
Plate 9, figures 5, 6

1884. *Cardiocarpus ovalis* LESQUEREUX, Coal Flora of Pennsylvania, v. 3, p. 810, pl. CIX, figs. 8, 9.

Description.—Small seed, slightly oblong, heart-shaped, 5–8 mm long, 5–6 mm wide, approximately 2 mm thick, flattened. Point of attachment of pedicel marked by a depression. No ring or margin surrounding the nucellus.

Remarks.—All of the specimens of this species were taken from the Route J collection site in Cedar County, Missouri. The specimens at hand compare well with those figured by Lesquereux (1884, p. 10, fig. 8), but are generally smaller. In all cases, the seeds were lacking marginal tissue. Lesquereux (1884, p. 811), in describing *C. conglobatus,* found the latter to have a rounder shape to the nucleus. In both species, the nucleus was often found separately, as in the present case.

CORDAITES CRASSINERVIS Heer
Plate 9, figure 7

1877. *Cordaites crassinervis* HEER, Flora Fossile Helvetiae, p. 150, pl. 30, fig. 4.

Description.—Leaf fragment 2 cm wide, venation pattern consisting of thick parallel veins, approximately 1 mm in diameter and 1 mm apart. No fine veins appearing between coarser veins.

Remarks.—There is a good comparison between the specimens at hand and those figured by Arnold (1949, p. 224, pl. xxx, fig. 4) and by Wood (1963, p. 67, pl. 12, fig. 5).

CORDAITES PRINCIPALIS (Germar) Geinitz
Plate 9, figure 8

1848. *Flabellaria principalis* GERMAR, Die Versteinerungen des Steinkohlenbirges von Weltin und Lobejun im Sallkreise, p. 55, pl. xxiii.

1855. *Cordaites principalis* GEINITZ, Die Versteinerungen des Steinkohlenformation in Sachsen, p. 41, pl. xxi, figs. 1, 2, 2a, 2b.

Description.—Cordaitean leaf with characteristic venation of from one to five fine veins between each pair of coarser veins.

Remarks.—No complete leaf specimens were found; the leaf apices were lacking in all cases. The leaf widths varied considerably, and because of this and the lack of apices, the identifications leave much to be desired. White (1899, p. 257) noted Grand'Eury's subdivisions of cordaitean leaf material and other additions by Renault and Zeiller and Sir William Dawson. Arnold (1949, p. 222) noted that the determination of cordaitean foliage was based on a rather variable set of characters—leaf outline combined with the arrangement of the coarse and fine striations on the lamina surface.

White (1899, p. 260, 261) described *C. communis* Lx. and *C. diversifolius* Lx., respectively. Both were found at the same collection site as of the present study. Lesquereux (1879, p. 534, 535) described both of the species mentioned above as having fine striations, one to four in number, between each pair of coarser striations. He further stated that the leaves of *C. diversifolius* are extremely variable. Although, on the basis of present-day classification of cordaitean leaves, the two species erected by Lesquereux and discussed by White would probably be grouped under *C. principalis*, I believe that this decision cannot be made until a large number of specimens of both species are brought together for a comparative study. Janssen (1939, p. 72) differentiated between *C. principalis* and

C. borassifolius on the basis of the leaf length, width, and apex shape, the latter being longer and wider. The apex is more rounded and obtuse in *C. principalis*. In both species, one to five fine veins occur between each pair of coarser veins. As Janssen pointed out, "when only small fragments of the leaves are found, it is frequently impossible to tell which of the two species is represented."

It is strongly evident that this group of foliage needs much work and revision.

DORYCORDAITES PALMAEFORMIS
(Goeppert) Zeiller
Plate 10, figure 1

1852. *Noeggerathia palmaeformis* GOEPPERT, Foss. Fl. d. Uebergansgeb., p. 216, pl. XV; pl. XVI, figs. 1–3.

1874. *Cordaites palmaeformis* WEISS, Foss. Fl. d. jungst. Steinkohl., p. 199, pl. XVIII, figs. 1, 2.

1877. *Cordaites (D o r y c o r d a i t e s) palmaeformis* GRAND'EURY, Flore carb. du dep. de la Loire, p. 214, pl. XVIII, figs. 4, 5.

1888. *Dorycordaites p a l m a e f o r m i s* (Goeppert) ZEILLER, Bassin houill. de Valenciennes, Description de la Flore Fossile (text), p. 632, 633; 1886 (Atlas) pl. XCIII, figs. 1, 2, 2a, 2b.

Description.—Portion of a straplike leaf. Parallel veins of equal or nearly equal size and close together, approximately 0.25 mm apart. Leaf apex lacking. One fragment measuring 45 cm long, width varying from 4.5–7.0 cm.

Remarks.—Janssen (1939, p. 71), in discussing cordaitean leaves and their placement in the three groups erected by Grand'Eury, remarked on the specific epithet *palmaeformis* under the genus *Dorycordaites* Goeppert because of the size and shape of the leaf. *Dorycordaites* included large leaves with pointed tips. Since tips are lacking in all of the specimens at hand, only the venation characteristic and width of the leaf were used to determine the taxonomic position.

The material at hand compares well with that figured by Zeiller (1888, pl. XCIII, figs. 1, 2, 2a) and labeled *Dorycordaites palmaeformis* Goeppert (sp.).

V
Comparison of This Flora with Other Pennsylvanian Floras

A Review of Previous Stratigraphic Studies Dealing with Missouri Fossil Floras

FIVE PREVIOUS STUDIES have been concerned with Missouri's Pennsylvanian Age floras. Four have dealt more or less extensively with the flora presently under study, in regard to stratigraphic position and geographic location. The fifth study described the flora found in the ironstone concretions at Windsor, Missouri; however, the Windsor flora is found in the shales above the Croweburg Coal (Cabaniss Subgroup) and is therefore younger than the flora under study.

Lesquereux (1884) provided a summary list of the fossil plants from the Coal Measures of Missouri. This comprised an enumeration of the species described or identified from Henry and Vernon counties up to that date. Of the 85 species listed by Lesquereux, 15 species were also found in the present study.

Hambach's study in 1890 (p. 83–84), like Lesquereux's, is of questionable use for comparative purposes because he omitted the stratigraphic level and exact location of collection. Hambach listed the species present, named the county in which they had been found, and noted that the specimens were from the Coal Measures. Hambach listed a total of 89 species; 23 comparable species were found during the course of the present study.

In 1893 White published a study of the flora of the southwestern Missouri basins located in Jasper and Lawrence counties, Missouri. Of the 34 species listed by White, 12 species are found in the Drywood flora. White also compared his flora with floras from Cannelton, Pennsylvania; Clinton, Missouri; Mazon Creek, Illinois; and the British provinces. He found that the greatest agreement was reached with the Clinton-Cannelton floras, but said that his flora appeared to be "rather younger" than the Clinton plants.

White, in 1897, published a report concerning the age of the lower coals of Henry County, Missouri. He noted that inferences of correlative aid must be cautiously drawn when dealing with fossil plants that are new or restricted to one region and when dealing with species having a wide vertical range.

White correlated the lower coals of Henry County with a stage later than the Morris, Brookville, and Clarion horizons in the Illinois, Ohio, and Pennsylvania Bituminous Series, but probably not extending above the Kittanning Group. When compared to the Anthracite Series, the fossil flora of Henry County compares with the fossil flora of the Marcy (D) Coal in the Northern Anthracite field. It was White's opinion that his flora was not older but was younger than the Middle Coal Measures of Great Britain and that the flora was presumably as late as the Transition Series. These series would be comparable to the present-day Westphalian and Staffordian series, respectively. White further qualified his correlation by stating that, possibly, the flora was comparable with the flora from the basal portion of the Upper Coal Measures or Radstockian Series.

White stated that the flora of the lower coals of Henry County and the flora of the Bully-Grenay zone of the Valencienne Basin of the Franco-Belgian coal field were nearly synchronous. He qualified his decision by stating that it was probable that the Missouri plant beds were slightly younger, since the Missouri pecopterid forms contained species of Stephanian character.

A study of the distribution of the Henry County fossil flora in other European basins allowed White to conclude that contemporaneous stages were found in the Geislautern beds near the top of the Saarbruck Series of the Rhenish coal regions, in the upper part of the Schatzler Series, and in the Radnitz Series in central Bohemia.

The study which has most importance in regard to the present study was that done in 1899 by White. White reported about 120 species, 30 of which were found during the course of this study. It is significant to note that there are 17 species in the present study of the Drywood Formation fossil flora which have not been previously reported in Hambach's, Lesquereux's, or White's lists of plants from that area. The following list of species new to Missouri eliminates the possibility of duplicate names and misinterpretations.

Aspidiaria sp. Presl
Knorria sp. Sternberg
Lycopodites sp. Brongniart
Annularia acicularis (Dawson) White
A. galioides (Lindley and Hutton) Kidston
Calamites cruciatus Sternberg

Calamostachys paniculata Weiss
Alethopteris davreuxi (Brongniart) Goeppert
A. decurrens (Artis) Zeiller
A. grandini (Brongniart) Goeppert
A. valida Boulay
Mariopteris speciosa Lesquereux
Neuropteris elacerata sp. nov.
Holcospermum Nathorst sp. ?
Artisia transversa (Artis) Sternberg
Cardiocarpus crassus Lesquereux
C. ovalis Lesquereux

White concluded that the Henry County flora was older than the plants of Mazon Creek and the Morris Coal in Illinois, and that they were very likely older than the Upper Kittanning Coal of Pennsylvania. In a comparison with plants of Great Britain, White decided that the Henry County flora was an intermediate between the Middle and Upper Coal measures.

Species of fossil plants from the Valencienne coal basin of France, studied by Zeiller (1888), has a remarkably high correlation with White's Missouri flora. More specifically, the Missouri flora had a very marked and preponderant affinity with the flora of the zone of Bully-Grenay. This was Zeiller's third or upper zone of the Westphalian. The presence in the Henry County flora of species more common in the Stephanian than in the Westphalian indicated to White that his flora was transitional between the upper Valencienne (Westphalian) and lower Stephanian.

White (1899) also made comparisons to the floras of Germany studied by Geinitz (1855, 1858), Stur (1875, 1877), and Feistmantel (1874). These comparisons provided further support for the assumption that the Missouri flora bore evidence of a transition flora between the Westphalian and Stephanian of Europe.

In 1915 David White published an account of the fossil floras of the Pennsylvanian in Missouri in which he reviewed and revised certain conclusions he had published in his U.S.G.S. Monograph 37 dealing with the age of the Henry County flora. White (1897) had correlated this coal flora with that of the Middle or Kittanning Group of coals in the Allegheny Formation. In the 1915 publication White concluded that the flora was Upper Pottsville because of the presence of such cheilanthites as *Eremopteris bilobata* and *Diplothmema obtusiloba*. Another very distinctly uppermost Pottsville indicator was *Mariopteris sphenopteroides*. White considered this taxon to be characteristic of the highest coal of the Kanawha Formation (Upper Pottsville), the Mercer Coal. White stated that *Neuropteris missouriensis* belonged unquestionably to the *N. flexuosa* group and that this group had no known representation above the Mercer coals in any American coal field.

Bode (1958) reported 19 species in the Croweburg Formation, of which seven species may be correlated with comparable species in the present fossil flora. Bode did not make any comparison to the older Missouri floras, but he did compare his specimens to the Mazon Creek concretion flora of Illinois. He stated that the Windsor flora was comparable to or older than the Mazon Creek flora. The presence of seven similar species between the present material and Bode's flora, which is younger, may be explained by the fact that the two formations, in which the floras occur, are not far apart within the same stratigraphic group and that the comparable species are wide ranging, stratigraphically.

Comparison with Fossil Floras of the United States

One of the most extensive studies of the American coal floras was conducted by Lesquereux (1880, 1884); however, Lesquereux erected so many new species, provided such terse descriptions, and used such a cumbersome lettering system for his coals, that correlation with other floras is of questionable quality.

White (1900) stated that the floras of the entire lower half of the Kanawha Formation were comparable with, if not actually older than, the Clarion and Kittanning floras, which follow them higher in the same section. When the fossil floras presented by White were compared to the present fossil flora of the Drywood Formation, the best correlation was made with the fossil flora occurring in the Kanawha Series. Seventeen fossil species from White's list of species found in the Kanawha Series were found also in this study in the Drywood Formation flora. Fourteen of White's species found in the Allegheny Series correlated with identical species in the present fossil flora. Since the number of identical species from the two series is so similar, it may be assumed that there is, within the Drywood flora, a representative transition fossil flora. These species, which transgress the time boundary, cannot be assigned a definite position within one of the two series discussed.

Of the 61 species of fossil plants found in the Drywood Formation and identified in this study, six similar species were also found in the Upper Lykens Coal flora. White (1900a) had listed 151 species of fossil plants as occurring within the Pottsville Formation in the type region of the southern anthracite field. None of the species identified for this study has been reported from collections from the Lower Lykens Coal group. For the most part, the six comparable fossil species in the Upper Lykens are those with extensive vertical ranges beginning in at least early Westphalian. The finding of six similar species does not indicate that the Drywood Formation is comparable in age to early Westphalian. Rather, the finding indicates that the plants persisted in the flora.

The fossil floras from the Arkansas coal fields were reported by White (1907) to be the same age as the Kansas Cherokee Formation flora and the Henry County, Missouri, basal Coal Measures flora. White's several coal floras were divided into groups and related to other floras. The Coal Hill Coal of Arkansas was placed at the level of the Brookville Coal in the Allegheny Formation in the Appalachian region. The Coal Ridge flora was considered by White to be the same age as, or older than, the Lower Kittanning Coal flora of the Allegheny Formation. The Paris Coal and its associated flora was placed as Upper Kittanning or younger.

Of the approximately 25 species listed as occurring in the flora of the Coal Hill Coal, 10 species occur in the flora under study. This flora, White concluded, represented a basal Alleghenian flora in which a few traces of Pottsville development could still be seen.

Seven species I found in the Drywood Formation flora were identified by White in the flora of the Coal Ridge Coal, which contained approximately 15 species. White pointed to an approximate contemporaneity between the Lower Coal in Henry County, Missouri, and the Coal Ridge flora. He doubted whether the flora was younger than the Lower or Middle Kittanning.

White's list of plants from the Paris flora of Arkansas contained only one species, of the total of seven, that is comparable to a species in the present study.

His list of 28 species of the Van Buren flora of Arkansas contained 12 species identical with those of the present study. The Van Buren flora was not considered by White to be younger than the Coal Ridge flora (basal Allegheny). I would place the Van Buren flora and the flora of the Drywood Formation as contemporaneous, although each contains some species different from the others. This might be attributed to the natural differences found in various collecting sites.

In 1913 White published *The Fossil Flora of West Virginia.* In this publication he listed species of fossil plants ranging from the Mississippian System to the Pleistocene Series.

The first occurrence of a comparable species in the West Virginia floras to that of the Drywood Formation is found in the Sewell Formation (Pennsylvanian, Pottsville Group). Of the 42 species listed, only one species, *Alethopteris serlii,* is found in the Drywood. White listed 24 species of fossil plants from the sandstone lentil of the Sewell Formation, and, of these, three species are in my collection of the flora of Missouri. In the Kanawha Formation (Pennsylvanian, Pottsville Group) of West Virginia, 114 species of fossil plants were found, and, of these, 18 species are in the current fossil flora. Of the 46 species listed from the Allegheny Formation fossil flora, 10 species are present in the Drywood Formation fossil flora. White listed 23 species from the Conemaugh Formation. Four of the species listed by White are in the flora under study. Only one of these was found in the Monongahela Formation (Pennsylvanian) of West Virginia. This was the wide-ranging *Neuropteris scheuchzeri*.

The foregoing comparisons illustrate that the best correlation of fossil floras between the Drywood Formation and a comparable formation in West Virginia is made with the Allegheny Formation.

Round (1926), in her study of the fossil flora of the Narragansett Basin of Rhode Island, made specific comparisons with White's 1899 flora of Henry County. Round found that over 50 per cent of the Rhode Island fossil plant species had been reported from Missouri by White. Of the 37 species listed by Round, 15 species are reported in this present study of the Missouri flora. There is a very close relationship between these two floras, even though they are so widely separated geographically.

A review of White's 1940 publication revealed no similarity between the fossil flora of the Stanley Shale and Jackfork Sandstone in the Morrow Series (southeast Oklahoma and west Arkansas) and the fossil flora of the Drywood Formation I collected in Missouri. White equated the floras of Oklahoma and Arkansas with the Culm of Europe, which is considered to be the Lower Coal Measures of England.

Read (1947) recognized nine successive floral zones in the Upper Paleozoic (numbered from the bottom up) in the eastern and Mid-Continent regions of North America. No representatives of Read's Zones 1 and 2 were found in this study. The Drywood Formation flora contains such elements of Read's Zone 3 as *Alethopteris decurrens* (not common), *Mariopteris speciosa* (common), *Cordaites principalis* (common), and *Sphenophyllum longifolium* (rare).

No plant species were found in the present study that could be related to the species assigned to Read's Zone 4.

The only elements of the Drywood flora that correspond to Read's Zone 5 are *Neuropteris scheuchzeri* and *Lepidophyllum* sp., both of which are common in this flora. However, as noted earlier, *N. scheuchzeri* is a wide-ranging species. Moreover, the identified *Lepidophyllum* sp. is not specific enough to use as a stratigraphic indicator. This zone was supposed to be characteristic of the major portion of the Kanawha Formation and, according to Read, also characteristic of the floras in the rocks in the Lampasas Group in the Mid-Continent region.

Read noted that the flora of Zone 6 occurred in the lower part of the Allegheny Formation in the Appalachian region and in the lower part of the Desmoines Group in the Mid-Continent region. In common with those species listed by Read as characteristic of Zone 6 were *Alethopteris serlii*, *N. scheuchzeri*, *Calamites suckowi*, *Annularia stellata*, and *Sphenophyllum emarginatum* from the flora under study.

Read characterized Zone 7 as "the Zone of *N. flexuosa* and the appearance of abundant *Pecopteris* species." This agrees somewhat with the flora of the present study. Read's Zone 7 occurs in the upper part of the Allegheny Formation or its equivalent, the upper part of the Desmoines Group. The extent of

Read's Zone 7 is questioned, since the specimens were recovered from the Drywood Formation, which is definitely considered to be Lower Desmoinesian.

Read's Zones 8 and 9 cannot be separated in the Mid-Continent region. They are represented in the Missouri flora by *Alethopteris grandini, N. scheuchzeri,* and *Annularia stellata.*

The closest correlation between the Drywood Formation flora and Read's zones is found in Zone 6. This is based on the species found in Zone 6 and on the general geologic range of each species.

As noted by Cridland et al. (1963, p. 60), Bode (1958) criticized Read's classification because it ignored some obvious similarities existing between European Carboniferous and American Pennsylvanian plant assemblages. Read made no comparison between European and American rocks based on plant fossils, nor did he make any comparison with previous European work on Upper Carboniferous classification. Cridland et al. also observed that Read stressed relatively less known American species of perhaps only local significance.

Although these criticisms were well founded, it should be noted that Read was one of the first paleobotanists to relate all of the floras of the United States.

Arnold (1949, p. 149) distinguished at least two and probably three floral zones within the Michigan Coal Basin. Of the 88 species listed by Arnold in his revised list of fossil plants occurring in Michigan, 18 occur also in this Missouri flora.

Of the distinctive species occurring in the Saginaw Coal—the "lower" flora, which is the lowest coal horizon in Michigan—only *Diplothmema obtusiloba* occurs in the Missouri flora at hand. Arnold's "intermediate" flora contains several representative species. Of these, only *Neuropteris scheuchzeri* occurs in the Drywood Formation flora. Arnold pointed out that there was an intergradation of the "intermediate" flora with both the "lower" and "upper" floras. The "upper" flora was characterized by the initial appearance of *Sphenophyllum emarginatum, Palmatopteris furcata, Annularia stellata,* and *A. sphenophylloides.*

Investigators have stated that the faunas of the Saginaw Group are Lampasan or Desmoinesian in age and equivalent to the Kanawha or Allegheny, respectively. Arnold correlated his Saginaw Coal flora with Read's Zone 3 and Dix's floral Zone D of South Wales. According to Arnold (1949, p. 153), the "intermediate" flora probably belonged to Read's Zone 5. The view was taken that the "upper" flora contained some Lower Allegheny entities but was, in general, probably not older than latest Kanawha. Arnold thought the "upper" flora was a transition flora to the Allegheny flora.

Some specimens are found in the present flora that are comparable to those in Arnold's "lower" and "intermediate" floras; however, in my opinion, the flora under study bears more resemblance to the "upper" flora, although it is probably slightly younger. Arnold (1949, p. 158) stated that the "upper" flora was not a true Allegheny flora, since the pecopterids that characterized the floras of Mazon Creek, Illinois, and Henry County, Missouri, were absent in his flora. He felt that such plants as *Neuropteris ovata, N. flexuosa, Alethopteris serlii, Pecopteris dentata,* and *P. pennaeformis* were typical Allegheny species, and these were lacking in the Michigan flora.

In commentary about the flora of the Michigan

Coal Basin, Arnold (1949, p. 152) noted that the study was made difficult by the extremely unsatisfactory and sketchy condition of the information relative to Pennsylvanian floras elsewhere in the eastern United States. He concluded that floras of many important coal horizons were unknown, that up-to-date descriptions of the Appalachian plant types were lacking, and that some species were not properly figured or illustrated. Arnold claimed that it was necessary to resort to the more ample and copiously illustrated literature on European coal floras, even though this method had the disadvantage of relying too heavily on foreign literature. Arnold felt there were many plant forms in the American Carboniferous that closely resembled contemporary European forms but which, if critically examined, would be recognized as specifically different. Further, Arnold concluded that comparison with European types would only tend to exaggerate differences between the floras of neighboring areas in the United States.

Cridland and others (1963) stated that the Pennsylvanian of North America and the Upper Carboniferous of Europe have many species in common and that their geologic succession is similar. They referred to Bode's conclusion that the differences which do exist between the floras of the two continents are no greater than the differences existing between several European coal basins.

Stewart (1950) listed 103 species in his report on the Carr and Daniels collections of fossil plants from Mazon Creek, Illinois. Twenty-two of the species in the Mazon Creek flora are also in the Drywood flora. Stratigraphically, the Mazon Creek flora is from a group higher in the column than the group in which the Drywood Formation flora is located. For this reason, it may be tentatively concluded that the two floras are not contemporaneous, although similar species are present in both. The extensive number of *Pecopteris* species in the Mazon Creek flora also suggests that it is younger than the flora under study.

In 1960 Read and Mamay published a paper dealing with the stratigraphic and lateral distribution of Upper Paleozoic megafossil floras in the United States. Although they divided the Mississippian, Pennsylvanian, and Permian systems into 15 floral zones, only Zones 9 (Mid-Pennsylvanian) and 10 (Upper Pennsylvanian) concern the present study. Zone 9 contained the characteristic plants *Pecopteris vestita, Mariopteris occidentalis, Neuropteris ovata, Linopteris rubella,* and cyatheoid pecopterids. This zone was related to the lower part of the Allegheny Formation in the Appalachian region, the upper part of the Sharp Mountain Conglomerate Member, and the Pottsville Formation of the southern anthracite field. It was also representative of the lower part of the Desmoines Series in the Mid-Continent region.

Zone 10, which is comparable to Read's Zone 7, was correlated with the lower part of the Conemaugh Formation and upper part of the Allegheny Formation of the Appalachian region and the upper part of the Desmoines Series in the Mid-Continent region.

As previously noted in the discussion of Read's zones, the closer correlation of the Drywood Formation is made with Read's Zone 6, comparable to Zone 9 of Read and Mamay. The flora of my study is transitional, fitting either Zone 9 or 10, but the position of the Drywood Formation stratigraphically indicates the placement in Read and Mamay's Zone 9.

In the Pennsylvanian flora of Kansas, Cridland, Morris, and Baxter (1963) found only *Alethopteris grandini* in the Krebs Formation. The rest of the specimens in their study came from younger formations. Seventy-eight species were listed by Cridland and others. Of the Kansas species, 19 are found also in the Missouri flora.

Cridland and others (1963, p. 61) correlated the Kansas Pennsylvanian plants from the Cabaniss Formation with those occurring in the upper part of the Westphalian and the Stephanian of Europe. Support for placement in Westphalian D was derived from the occurrence of *Alethopteris serlii, Neuropteris ovata, N. scheuchzeri,* and *Sphenopteris obtusiloba* in the Kansas flora. These species are not commonly found above Westphalian D in Europe. It was noted that these species do in fact occur in older rocks. Other species found in younger rocks were related to Stephanian Age plants in Europe.

The Kansas flora is definitely younger than the Missouri flora under study, although there are common species represented in both. These should be interpreted as species with long ranges that are not reliable as indicators of a specific age.

Cridland and others (1963, p. 64) stated that the plants from the Kansas Cabaniss Formation agreed fairly well with those White (1899) described from the Cherokee Group near Clinton, Missouri, and with those White (1893) described from southwestern Missouri. Agreement of floras was found also with the Windsor, Missouri, ironstone concretion assemblage, studied by Bode (1958). This flora occurs in the shale above the Croweburg Coal. This assemblage, occurring above the Drywood Formation in Missouri, was found to have Stephanian Age plants as well as Westphalian. Cridland and others suggested that the Windsor plants are from the upper part of the Westphalian D.

In the 1956 Correlation Chart compiled by Siever and found in Wanless' (1956) *Classification of the Pennsylvanian Rocks of Illinois as of 1956,* the Drywood Formation of Missouri is correlated with the Tradewater Group of Illinois and with the Pottsville Series of Indiana. More specifically, this correlates with the range of the Lower and Upper Block coals of Indiana from which Wood (1963) obtained the Stanley Cemetery flora.

A comparison of the Drywood fossil flora with the Lower Block Coal flora of Indiana (Wood, 1963) revealed 25 species in common. Wood found that the Indiana flora consisted of 86 species. There is a very close relationship between these two floras, because both possess fossil species that are indicative of transitional stages. Such species as *Diplothmema furcatum* and *D. obtusiloba* are indicators of a more primitive fern flora than the pecopterids, which also appear in both floras. It would appear that there is a closer correlation between the two floras than is evident in comparison of numbers of species. Wood compared the Indiana flora to White's (1899) flora of Henry County, but stated that his flora was more closely similar to that of the Michigan Coal Basin.

On the basis of the foregoing discussions, it may be concluded that the Drywood Formation fossil assemblage contains both Kanawhan and Alleghenian taxa.

Comparison with Fossil Floras of Canada

Marie Stopes (1914) found that of the over 80 species described periodically from the "Fern Ledges" flora of St. John, only about 40 were of value in comparing her flora with others. Of the 41 species listed by Stopes, only 10 species were identified in the Drywood flora. In another list of principal species—28 species listed—Stopes reported five species that were identical to those reported by White (1899) from Missouri beds, and two species that were exceedingly similar if not identical. Stopes concluded that the Fern Ledge flora corresponded to the lowest zone of the Middle Westphalian.

Bell (1938, 1940, 1944, 1961) has provided scientists with almost all of their recent knowledge of the Canadian fossil floras.

In 1938 Bell divided the Morien Series of the Sydney coalfield, Nova Scotia, into three zones. The zones, in ascending order, were: Zone of *Lonchopteris eschweileriana* and of *Alethopteris lonchitica*, Zone of *Linopteris obliqua*, and Zone of *Ptychocarpus unitas*.

The *Lonchopteris* zone contained 24 species, nine of which are found in the Drywood Formation. Bell related this zone to Westphalian B and the lower Westphalian C, since those species most characteristic of Westphalian C—for example, *Neuropteris scheuchzeri* and *Alethopteris serlii*—were few in number.

Bell's *Linopteris obliqua* zone was comprised of 88 species. At least 16 species are identical to species in the present Missouri study. This zone, Bell concluded, comprised a flora characteristic of Yorkian species intermixed with species characteristic of the Radstockian. It was, consequently, Staffordian in age. Such species as *Neuropteris scheuchzeri, N. ovata, N. heterophylla, Alethopteris serlii, Annularia sphenophylloides, A. stellata,* and *Sphenophyllum emarginatum* range up into the Radstockian. Several of the species listed above, such as *N. scheuchzeri, N. heterophylla, Annularia sphenophylloides,* and *A. stellata,* are common in the Drywood flora. Bell compared the *Linopteris obliqua* zone with Zeiller's Flora C of northern France. He also stated that this zone is approximately the same age as the flora of Henry County, Missouri.

Bell's third zone, the *Ptychocarpus unitas* zone, contained 84 species. The present flora contains 17 species found in the *Ptychocarpus* zone, which Bell concluded was comparable to the Radstockian. Bell found a close affinity existing between the flora of the *Ptychocarpus unitas* zone and the uppermost flora of the McAlester coalfield, Oklahoma. He further stated that the Kittanning Group of the Allegheny Series was equivalent in age to either an upper part of the *Linopteris obliqua* zone or a basal part of the *Ptychocarpus unitas* zone.

The greatest degree of similarity was found between the flora of the lowest zone in Nova Scotia and the Drywood Formation flora, but most of these were not common in the Missouri flora. Only *Annularia sphenophylloides* might be considered common; the rest were infrequent in their occurrence. Therefore, a conclusion in agreement with Bell may be drawn. The Drywood Formation fossil flora correlates with the *Linopteris obliqua* zone of Nova Scotia.

In his study of the Pictou coalfield, Bell (1940) divided the Pennsylvanian into four series: Canso (Namurian), Cumberland (Westphalian lower B), Stellarton (Westphalian C and Westphalian upper B), and Pictou (?Westphalian upper B).

The Canso Series included fresh-water strata that directly overlay Mississippian Age strata and contained only three species, none of which was found in this Missouri flora. Only seven species were listed for the Cumberland Series, and, of these, only *Cordaites principalis* is represented in this flora.

The Stellarton Series of the Pictou coalfield was divided into two divisions. Bell found no fossils diagnostic of age in the lower division but did mention *Cordaites principalis, Cardiocarpon, Neuropteris, Sigillaria, Samaropsis cornuta,* and a single pinnule of *Alethopteris grandini.* This is insufficient evidence for a comparison. On the other hand, Division II of the Stellarton contained 46 species, of which nine are in the flora under study. Of the three members making up Division II, eight of the nine species comparable with this Drywood flora were found in the Thorburn Member. Bell concluded that the Thorburn Member was of Westphalian C Age because of such characteristic species as *Neuropteris scheuchzeri, Alethopteris serlii,* and *Sphenophyllum emarginatum.* Furthermore, Bell concluded that the Thorburn Member was probably contemporaneous with a part of the *Linopteris obliqua* zone of the Morien Series of the Sydney coalfield. Of the six species listed by Bell (1940, p. 51) for the Pictou Series, *Annularia sphenophylloides* and *Sphenophyllum emarginatum* are also present in the Drywood Formation. Bell indicated that the Pictou Series and the Stellarton were of similar age.

Bell's (1944) study of the Carboniferous rocks and fossil floras of northern Nova Scotia listed species from the Riversdale, Cumberland, and Pictou groups. Bell found 50 species in the Riversdale Group, and, of these, only seven are in this Missouri flora. Bell listed 82 species from the Cumberland Group. Of the 11 species also occurring in the present flora, only two species were considered by Bell to be confined to the

Cumberland Group. As in Bell's 1938 study, the best correlation between the Drywood Formation flora and a Nova Scotian flora was made with the Pictou Group. In the Pictou Group was found the zone of *Linopteris obliqua*. Bell concluded that the age of the Pictou Group in the northern Nova Scotia area was mainly Westphalian C and D.

The most recent of Bell's studies was published in 1962 and dealt with the flora of the Pennsylvanian Pictou Group of New Brunswick. Of the 65 species listed by Bell as occurring in the New Brunswick flora, 14 species occur in the Drywood Formation flora. Of these 14 species, three were found in only the upper *Linopteris obliqua* zone, four occurred only in the lower portion of the *Linopteris obliqua* zone, and seven species occurred in both the upper and lower zones. Bell correlated the Pictou Group of New Brunswick with the Westphalian C. Again, the presence of such species as *Neuropteris scheuchzeri, Sphenophyllum emarginatum, Annularia sphenophylloides,* and *Asolanus camptotaenia* gave support to Bell's premise that his flora was of Westphalian C Age. The findings of the present study support Bell's conclusions and concur that the Canadian floras of the *Linopteris obliqua* zone are, on the whole, comparable to the flora found in the Henry County, Missouri, area.

Comparison with Fossil Floras of Europe

In 1888 Zeiller published his excellent study of the Valencienne Coal Basin and the fossil flora located there. Zeiller found a total of 166 species. He divided the flora of the basin into three divisions and grouped under each division the general collection area indicating whether it was considered to be lower, middle, or upper zone. When the Drywood Formation fossil flora was compared to Zeiller's fossil flora, seven Missouri species were found in common with Zeiller's lower zone; 18 Drywood fossil species were found in Zeiller's middle zone; 21 Missouri species were found in the upper zone flora. The highest degree of correlation between the Drywood fossil flora and those of Zeiller's Valencienne Coal Basin may be drawn with Zeiller's upper zone.

Of the 61 species identified in the present study of the Drywood Formation fossil flora, only six species are also found in Kidston's (1923–25) *Fossil Plants of the Carboniferous Rocks of Great Britain.* One species, *Diplothmema furcatum* ranges from Lanarkian to Westphalian deposits. Two species, *Sphenopteris mixta* and *Mariopteris speciosa,* are confined to Westphalian deposits. *Sphenopteris obtusiloba* ranges from Lanarkian to Staffordian strata, while *M. sphenopteroides* is confined to rocks of Staffordian Age. *Pecopteris dentata* was found in strata ranging in age from Lanarkian to Radstockian.

Only a very general comparison can be drawn, because of the limited number of similar species found between the two fossil floras, but it would seem that the Drywood fossil flora has a greater similarity to fossil plants from the Westphalian of England than to those of older or younger age.

There can be no correlations drawn between the Drywood flora in Missouri and the coal ball flora described by Koopmans (1928) from the Westphalian A of the Netherlands. Each of the 61 species in the Drywood fossil flora was different than the 68 species in the Koopmans flora.

In 1933, Dix proposed the establishment of nine zones for the Upper Carboniferous rocks in Wales. She further related the broad plant zones of Wales to those Bertrand found in northern France and to the floral divisions identified in Holland by Jongmans, in Belgium by Renier, in Westphalia by Gothan and others, and in the Donetz coalfield by Zalessky.

Only six species are common between the Drywood Formation and the 33 species listed under Dix's Flora C. Of Dix's Flora D, which contains 40 species, eight are found in this flora. Flora E contained 106 species, 15 of which are found in the Drywood Formation fossil flora. Flora F, which contained 79 species, compared with only 11 species in my flora. Fourteen of the species found in the Missouri flora were also found among the 82 species of Flora G. Eleven of the Dix's 73 species of Flora H were found in the Drywood flora. The uppermost zone, Flora I, contained 69 species, but only nine of them are found in this flora.

The closest correlations between the floras reported by Dix and this Missouri flora are in Floras C, D, and G—they are 18, 20, and 17 per cent, respectively. The entire range of these floras would extend from the lower Westphalian (Yorkian) to the Middle of the Stephanian.

There are 127 species listed in the summary chart, presented by Gothan and Remy (1957), of plants found often in the Paleozoic of Germany. In addition to the listing of these plants, a vertical range was given for each. Of the 127 fossil species reported by Gothan and Remy, 20 fossil species were limited to ranges completely outside of the Westphalian-Stephanian ages; that is, they occur in the Dinant and Rotliegend strata. These species were not considered in the comparison of floras. When the 21 species that occur in the Drywood Formation fossil flora and in the German flora are compared, in respect to their corresponding ranges in Europe, 19 of the 21 species are found to have Westphalian C Age in common.

Wagner (1962) published an account of the floral succession of the Carboniferous in northwest Spain. This succession ranged from the Visean to the Autunian. Of the nine ages presented, the closest correlation of floras between the Drywood Formation and Spanish Carboniferous is made with that flora which Wagner considered to be Westphalian D. Of the 31 Westphalian D species listed by Wagner, six occur in the Missouri flora, but these are generally long-ranging species and thus are not as significant in determining age. The second closest relationship is found with the flora Wagner considered to be Stephanian A. This flora contained 47 species, of which five species are found here. There appears to be a higher degree of correlation with floras of younger age in northwest Spain than with what is usually considered to be correlative floras of the more northerly European countries. The significance of this correlation is not known at the present time.

In the 1944 publication by Moore and others, *Correlation of Pennsylvanian Formations of North America,* the Missouri column is blank in the range where the Drywood Formation should be located. But, from what is known of the present-day position of the formation and according to Moore and others, the Drywood Formation should be correlated with Westphalian C, the Flammkohle of Germany, the Staffordian of England, Dix's Zone of *Neuropteris rarinervis,* and Read's Zone 6—Zone of *Neuropteris rarinervis*. An actual correlation of floras has shown such a generalized correlation to be misleading, however, in some cases.

Comparison with Fossil Floras of North Africa

Jongmans (1953) published a report of North African Paleozoic floras and their correlations with floras of other regions of the world. Of the several locations in Algeria and Morocco and their associated floras, the closest correlations to the Missouri flora are with the floras of the Kenada Basin, Djebel Mezarif, and Sidi Brahim. The flora of the Kenada Basin contained 28 species, eight of which were found in the Drywood Formation. This flora was considered to be of Westphalian C Age. The Djebel Mezarif flora contained 34 species. Ten comparable species were found in Missouri. Jongmans considered this flora to be upper Westphalian C and possibly lower Westphalian D. The best comparison is found with the flora of Sidi Brahim, which contained 13 species, four of which were also found in the present flora. This flora was considered to be Westphalian B.

When interpreting the importance of the correlations, not much emphasis is placed on the greater similarity to the Sidi Brahim flora. The percentages of the Drywood flora as compared to the African floras are all very similar: Kenada Basin (28.5 per cent), Djebel Mezarif (29.4 per cent), and Sidi Brahim (30.7 per cent), respectively. Such a close percentage comparison with a succession of ages lends support to the conclusion that the Drywood fossil flora is a transitional type.

VI

Conclusions

A COMPARISON OF THE FLORA from the Drywood Formation of Missouri with the fossil floras in other areas of the United States, Canada, England, Wales, France, Germany, Spain, and North Africa reveals that the flora under study is possibly of Upper Kanawhan Age, Upper Pottsville Age, or Lower Allegheny Age and would be correlated with floras ranging from Westphalian B to Staffordian. The closest correlation is with the Westphalian C stage.

In this assemblage, a large number of specimens of *Annularia* are found that intergrade between *A. acicularis,* which is the older species (Lower Pottsville to Lower Allegheny per Abbott, 1958, p. 306) and *A. stellata,* which is the younger species (Pottsville to Dunkard per Abbott, 1958, p. 322). The intergradation suggests that a shift from the older species to the younger has occurred in this flora. The shift would be indicative of a transitional period.

The presence of such species as *Asterophyllites equisetiformis, Diplothmema furcatum, Alethopteris decurrens,* and *Cordaites principalis* indicates a Kanawhan age for a portion of the flora.

The presence of such species as *Alethopteris grandini, A. serlii, Neuropteris ovata* forma *flexuosa,* and *Pecopteris dentata* indicates an Alleghenian or Desmoinesian age for another portion of the flora.

Because such a wide age span is indicated, it is difficult to date, by paleobotanical means alone, the strata within which the flora is located. However, the geologic data, in conjunction with the paleobotanical evidence, indicate that certain species are more dependable as age determinants in this particular area. The more dependable species are *Alethopteris serlii, Neuropteris ovata* forma *flexuosa,* and *Pecopteris dentata.*

The study demonstrated the advisability of gathering many collections from several sites in the same stratum because of the limitations of any single site. At one site, only lepidodendrid material was found; whereas, at another site, only seeds had been preserved. It is clear, from this study and others, that species of plant megafossils are not equally distributed throughout their lateral range and that, unless many collections are made, the flora may contain a bias or be lacking certain species completely. This limitation of individual sites may be due to ecological factors. In addition, much depends on each individual investigator as to whether his flora is large or small, whether it has many new species or none, whether the descriptions and illustrations he publishes are good or poor. Thus, the value of the flora represents the care with which the researcher has approached the problem.

The major components of the flora of the Drywood Formation are herbaceous and arborescent Lycopods and Calamites, sphenophylls, ferns, seed ferns, and gymnosperms of the Order Cordaitales. Such a generalized flora appears to compare favorably with Pennsylvanian or Upper Carboniferous floras found elsewhere. The absence of *Sigillaria* and *Lepidodendron* is noteworthy.

VII

Summary

A STUDY of the Drywood Formation in Henry and Cedar counties, Missouri, revealed a flora of 61 species of fossil plants. These macrofossils contained representatives of the orders Lepidodendrales, Equisetales, Sphenophyllales, Filcales, Cycadofilicales, and Cordaitales. The macrofossils were examined, described on the basis of the specimens found, and representative material of each species illustrated.

The geology of the depositional area is discussed and is shown to have a strong effect upon the continuity of the formation, the paleoecology of the area, and the enclosed plant fossils.

One new species is described in *Neuropteris;* one new combination is made in *Lepidostrobophyllum*. Of the 61 species in 29 genera, 21 were not previously described from this area.

In addition to increasing the lateral extent of the flora, redefining the species involved, and increasing the number of species found in the floral assemblage of the Drywood Formation, comparison is made with Lower to Middle Pennsylvanian floras from eight states in the United States, two Canadian provinces, England, Wales, France, Netherlands, Belgium, Germany, Spain, and Morocco and Algeria.

List of References

ABBOTT, M. L. 1958. The American species of *Asterophyllites, Annularia,* and *Sphenophyllum.* Amer. Paleontol., Bull. 38:289–368.

———. 1963. Lycopod fructifications from the upper Freeport (no. 7) Coal in southeastern Ohio. Palaeontographica Abt. B. 112:93–118.

ABERNATHY, G. E. 1938. Cyclical sedimentation in the Cherokee. Kansas Acad. Sci., Trans. 41:193–197.

ADAMS, G. E., G. H. GIRTY, and D. WHITE. 1903. Stratigraphy and paleontology of the Upper Carboniferous rocks of the Kansas section. U. S. Geol. Surv., Bull. 211:1–123.

ANDREWS, E. B. 1875a. Description of fossil plants from the Coal Measures of Ohio. Ohio Geol. Surv. Paleontol., Rept. 2:413–426.

———. 1875b. Notice of new and interesting coal plants. Amer. Jour. Sci. & Arts, ser. 3, 10:462–466.

ARNOLD, C. A. 1934. A preliminary study of the fossil floras of the Michigan coal basin. Michigan University, Museum Paleontol., Contr. 4:177–204.

———. 1947. An introduction to paleobotany. 433 p. New York: McGraw-Hill Book Company, Inc.

———. 1949. Fossil flora of the Michigan coal basin. Michigan University, Museum Paleontol., Contr. 7:131–269.

———. 1953. Fossil plants of early Pennsylvanian types from central Oregon. Palaeontographica Abt. B. 93:61–68.

BAKER, D. A. 1962. Geology of the Stockton Quadrangle, Cedar County, Missouri. Unpublished Master's thesis, University of Missouri.

BAXTER, R. W. 1952a. The coal age flora of Kansas. I. *Microspermopteris aphyllum* var. *kansensis* var. nov. Kansas Acad. Sci., Trans. 55:101–103.

———. 1952b. The coal age flora of Kansas. II. On the relationships among the genera *Etapteris, Scleropteris* and *Botrychioxylon.* Amer. Jour. Bot. 39:263–274.

———. 1953a. The coal age flora of Kansas. III. *Psaronius cooksonii:* a new species showing *Caulopteris* features. Bot. Gaz. 115:35–44.

———, and E. A. ROTH. 1953b. The coal age flora of Kansas. IV. *Calamopitys kansanum,* a new species from the Pennsylvanian of Kansas. Kansas Acad. Sci., Trans. 56:220–226.

———, and E. L. HARTMAN. 1954. The coal age flora of Kansas. V. A fossil coniferophyte wood. Phytomorphology 4:316–325.

BELL, W. A. 1938. Fossil flora of Sydney Coalfield, Nova Scotia. Canada Geol. Surv., Mem. 215, 334 p.
———. 1940. The Pictou Coalfield, Nova Scotia. Canada Geol. Surv., Mem. 225, 161 p.
———. 1944. Carboniferous rocks and fossil floras of northern Nova Scotia. Canada Geol. Surv., Mem. 238, 277 p.
———. 1962. Flora of Pennsylvanian Pictou group of New Brunswick. Canada Geol. Surv., Bull. 87, 71 p.
BERTRAND, P. 1932. Bassin houiller de la Saar et de la Lorraine. I. Flore fossile. Études des Gîtes Min. de la France Fasc. 2, Aléthoptéridees: 61–107.
BODE, H. 1958. Die floristische Gliederung des Oberkarbons der Vereinigten Staaten von Nordamerika. Z. Dt. Geol. Ges. 110: 217–259.
BOLTON, E. 1962. A critical study of certain species of the genus *Neuropteris* Brongniart. London. Jour. Linnean Soc. Bot. 47:295–327.
BROADHEAD, G. C. 1868. Coal Measures in Missouri. St. Louis Acad. Sci., Trans. 2:311–333.
———. 1873. Preliminary report on the iron ores and coal fields, from field work of 1872, Part II of geology of northwestern Missouri. Missouri Geol. Surv. 441 p.
———. 1895. The Coal Measures of Missouri. Missouri Geol. Surv. 8:353–395.
BUNBURY, C. T. F. 1847. Descriptions of fossil plants from the coal field near Richmond, Virginia. London. Quart. Jour. Geol. Soc. 3:281–288.
CANRIGHT, J. E. 1959. Fossil plants of Indiana. Indiana Dept. Cons. Geol. Surv., Rept. Prog. 14, 45 p.
CONDIT, C., and A. K. MILLER. 1951. Concretions from Iowa like those of Mazon Creek, Illinois. Jour. Geol. 59:395–396.
CRIDLAND, A. A., J. E. MORRIS, and R. W. BAXTER. 1963. The Pennsylvanian plants of Kansas and their stratigraphic significance. Palaeontographica Abt. B. 112:58–92.
CROOKALL, R. 1955. Fossil plants of the Carboniferous rocks of Great Britain (Sec. Sect.). Great Britain Geol. Surv. Paleontol., Mem. 4:1–84.
———. 1959. Fossil plants of the Carboniferous rocks of Great Britain (Sec. Sect.). Great Britain Geol. Surv. Paleontol., Mem. 4:85–216.
DARRAH, W. C. 1936. American Carboniferous floras. Deuxième Congrès pour l'avancement des Études de Stratigraphie et de Géologie du Carbonifère, Compte Rendu 1:109–129.
DUNBAR, C. E., and J. RODGERS. 1957. Principles of stratigraphy. 356 p. New York: John Wiley and Sons, Inc.
ELIAS, M. K. 1931. On a seed-bearing *Annularia* and on *Annularia* foliage. Kansas University Sci. Bull. 20:115–159.

―――. 1936. Character and significance of the late Paleozoic flora at Garnett. Jour. Geol. 44:9–23.

―――. 1937. Elements of the Stephanian flora in the mid-continent of North America. Deuxième Congrès pour l'avancement des Études de Stratigraphie et de Géologie du Carbonifère, Compte Rendu 1:203–212.

FONTAINE, W. F., and I. C. WHITE. 1880. The Permian or Upper Carboniferous flora of West Virginia and southwestern Pennsylvania. Pennsylvania, Second Geol. Surv., Rept. PP, 134 p.

FOSTER, J. W. 1853. New species of fossil plants from Ohio. Ann. Sci. 1:128–129.

GOTHAN, W., and W. REMY. 1957. Steinkohlenpflanzen. 247 p. Essen.

GRANGER, E. 1820. Notice of vegetable impressions on rocks connected with the coal formation of Zanesville (Ohio). Amer. Jour. Sci. 3:5–7.

GREENE, F. C., and W. F. POND. 1926. The geology of Vernon County. Missouri Bur. Geology and Mines, ser. 2, 19:21–141.

HAMBACH, G. 1890. A preliminary catalogue of the fossils occurring in Missouri. Missouri Geol. Surv. Bull. 1:60–85.

HARLAN, R. 1831. Notice of fossil vegetable remains from the Bituminous Coal Measures of Pennsylvania. Pennsylvania Geol. Surv., Trans. 1:1–256.

HAWORTH, E., and M. KIRK. 1894. The Neosho River section. Kansas University. Quart. 2:104–112.

HENDRICKS, T. A., and C. B. READ. 1934. Correlations of Pennsylvanian strata in Arkansas and Oklahoma coal fields. Amer. Assoc. Petroleum Geologists, Bull. 18:1050–1058.

HILDRETH, S. P. 1836. On the bituminous coal deposits of the Ohio Valley, with notes on fossil organic remains. Amer. Jour. Sci., ser. 1, 29.

HINDS, H. 1912. Coal deposits of Missouri. Missouri Bur. Geology and Mines, ser. 2, 2:1–503.

―――, and F. C. GREENE. 1915. The stratigraphy of the Pennsylvanian series in Missouri. Missouri Bur. Geology and Mines, ser. 2, 13:1–407.

HOWE, W. B. 1956. Stratigraphy of pre-Marmaton Desmoinesian (Cherokee) rocks in southeastern Kansas. Kansas State Geol. Surv., Bull. 123, 132 p.

HOWE, W. B., and J. W. KOENIG. 1961. The stratigraphic succession in Missouri. Missouri Div. Geol. Surv. and Water Resources, Repts., Ser. 2, 40, 185 p.

JACKSON, T. F. 1916. The description and stratigraphic relationships of fossil plants from the Lower Pennsylvanian rocks of Indiana. Indiana Acad. Sci., Proc. 26:405–439.

JANSSEN, R. E. 1937. A key for the identification of plant impressions from the Middle Pennsylvanian of Illinois. Unpublished

Master's thesis, University of Chicago.

———. 1939. Leaves and stems from fossil forests: A handbook of the paleobotanical collections in the Illinois State Museum. 190 p. Springfield, Illinois: State of Illinois, Division of the Illinois State Museum.

———. 1940. Some fossil plant types of Illinois. Illinois State Museum, Sci. Paper 1, 116 p.

JONGMANS, W. J. 1911. Anleitung zur bestimmung der Karbonpflanzen West-Europas mit besonderer berucksichtigung der in den Niederlanden und den Benachbarten Landen sefunden oder nach zu erwartenden arten. Meded. Rijks Opsporing v. Delfstoffen 1 (3), 482 p. 's-Gravenhage.

———. 1913. Fossilium catalogus II: Plantae. Pars 1, in W. J. Jongmans. (Ed.) Fossilium catalogus II: Plantae. 's-Gravenhage: Uitgeverij Dr. W. Junk, 1–52.

———. 1930. Fossilium catalogus II: Plantae. Pars 16, in W. J. Jongmans. (Ed.) Fossilium catalogus II: Plantae. 's-Gravenhage: Uitgeverij Dr. W. Junk, 329–650.

———, and W. GOTHAN. 1934. Florenfolge und vergleichende stratigraphie des karbons der östlichen Staaten Nordamerikas. Vergleich mit Westeuropa. Jaarverslag Geol. Bur. Heerlen, 17–44.

———. 1953. Le Paléozoique Nord-Africain et ses corrélations avec celui des autres régions du monde. Congrès Geologique International, C. R. de la dixneuvieme session, 1952, sec. 2, Fasc. 2:49–64.

———, and S. J. DIJKSTRA. 1958. Fossilium catalogus II: Plantae. Pars 34, in S. J. Dijkstra. (Ed.) Fossilium catalogus II: Plantae. 's-Gravenhage: Uitgeverij Dr. W. Junk, 7:463–512.

———, and S. J. DIJKSTRA. 1960a. Fossilium catalogus II: Plantae. Pars 40, in S. J. Dijkstra. (Ed.) Fossilium catalogus II: Plantae. 's-Gravenhage: Uitgeverij Dr. W. Junk, 13:1039–1168.

———, and S. J. DIJKSTRA. 1960b. Fossilium catalogus II: Plantae. Pars 41, in S. J. Dijkstra. (Ed.) Fossilium catalogus II: Plantae. 's-Gravenhage: Uitgeverij Dr. W. Junk, 14:1169–1258.

———, and S. J. DIJKSTRA. 1960c. Fossilium catalogus II: Plantae. Pars 44, in S. J. Dijkstra. (Ed.) Fossilium catalogus II: Plantae. 's-Gravenhage: Uitgeverij Dr. W. Junk, 17:1453–1596.

———, and S. J. DIJKSTRA. 1962a. Fossilium catalogus II: Plantae. Pars 50, in S. J. Dijkstra. (Ed.) Fossilium catalogus II: Plantae. 's-Gravenhage: Uitgeverij Dr. W. Junk, 23:2083–2160.

———, and S. J. DIJKSTRA. 1962b. Fossilium catalogus II: Plantae. Pars 52, in S. J. Dijkstra. (Ed.) Fossilium catalogus II: Plantae. 's-Gravenhage: Uitgeverij Dr. W. Junk, 25:2259–2348.

KIDSTON, R. 1893. On the fossil plants of the Kilmarnock, Galston, and Kilwinning coal fields, Ayrshire. Part II. Edinburgh. Roy. Soc., Trans. 37 (16): 307–359.

———. 1895. On *Lepidophloios,* and on the British species of the genus. Part III. Edinburgh. Roy. Soc., Trans. 37 (25): 529–563.

———, and W. JONGMANS. 1917. A monograph of the *Calamites* of western Europe, in W. J. Jongmans, Flora of the Carboniferous of the Netherlands and adjacent regions. Meded. Rijks Opsporing van Delfstoffen 1 (7): 207p 's-Gravenhage.

———. 1923–25. Fossil plants of the Carboniferous rocks of Great Britain. Parts 1–6. Great Britain Geol. Surv. Paleontol., Mem. 2, 681 p.

KOOPMANS, R. G. 1928. Researches on the flora of the coal-balls from the "Finefrau-Nebenbank" Horizon in the Province of Limburg (The Netherlands) in W. S. Jongmans, Flora en fauna van het Nederlandsche Karboon. I. Nederlandsche Mijngebied Geol. Bur. 53 p.

LANGFORD, G. 1958. The Wilmington coal flora from a Pennsylvanian deposit in Will County, Illinois. 361 p. Downer's Grove, Illinois: Esconi Associates.

LESLEY, J. P. 1889–90. A dictionary of the fossils of Pennsylvania and neighboring states named in the reports and catalogue of the survey. Pennsylvania, Second Geol. Surv., Rept. P4, 3 vols.

LESQUEREUX, L. 1854. New species of fossil plants from the anthracite and bituminous coal fields of Pennsylvania. Biol. Jour. Nat. Hist. 6:409–431.

———. 1857. Palaeontological report of the fossil flora of the Coal Measures of the western Kentucky coal field. Kentucky Geol. Surv., Third Rept. 3:499–556.

———. 1858. Fossil plants of the coal strata of Pennsylvania. Pennsylvania, Geology of. 2(2):835–884.

———. 1860. Botanical and palaeontological report on the geological survey of Arkansas. Arkansas, Second Report on the Geol. 308–317.

———. 1861. Report of the fossil flora and on the stratigraphical distribution of the coal in the Kentucky coal fields. Kentucky Geol. Surv., Fourth Rept. 4:331–437.

———. 1866. Enumeration of the fossil plants found in the Coal Measures of Illinois, with descriptions of new species. Illinois Geol. Surv. 2:427–470.

———. 1869. Fossil plants of the Main Sewanee Jackson Coal, etc. of Tennessee. Tennessee, Geology of. 408–409.

———. 1870. Report on the fossil plants of the Illinois coal fields. Illinois Geol. Surv. 4:375–508.

———. 1875. Partial list of coal plants from the Alabama fields and discussion of the geological positions of several coal seams. Alabama Geol. Surv., Progress Rept. 75–82.

———. 1876. On species of marine plants from the Carboniferous Measures. Indiana Geol. Surv., Seventh Ann. Rept. 134–145.

———. 1878. On *Cordaites* and their generic divisions in the Carboniferous formations of the United States. Philadelphia. Amer. Philosoph. Soc. 17:315–335.

———. 1879. On a branch of *Cordaites,* bearing fruit. Philadelphia. Amer. Philosoph. Soc.

———. 1880. Description of the coal flora of the Carboniferous formations in Pennsylvania and throughout the United States. Pennsylvania, Second Geol. Surv., Rept. Progress P. 1:1–354, 2:355–694, Atlas, 1879.

———. 1884. Description of the coal flora of the Carboniferous formation in Pennsylvania and throughout the United States. Pennsylvania, Second Geol. Surv., Rept. Progress P. 3:695–977.

———. 1887. On the character and distribution of Paleozoic plants. Pennsylvania, Geol. Surv., Rept. for 1886, part 1: 457–522.

LYELL, C., and C. T. F. BUNBURY. 1846. Observations on the fossil plants of the coal field of Tuscaloosa, Alabama. Amer. Jour. Sci. & Arts, ser. 2, 2:228–233.

MACKNIGHT, F. C. 1938. The flora of the Grape Creek Coal at Danville, Illinois. Unpublished Ph.D. thesis, University of Chicago.

MAMAY, S. H., and C. B. READ. 1956. Additions to the flora of the Spotted Ridge Formation in central Oregon. U. S. Geol. Surv., Prof. Paper 274-I:211–226.

MARBUT, C. F. 1898. Geological descriptions of the Clinton, Calhoun, Lexington and Richmond sheets. Missouri Geol. Surv. 12:20–308.

MEEK, F. B. 1875. Descriptions of new species of fossil plants from Allegheny County, Virginia. Washington, Philosoph. Soc., Bull. 2:26–44.

MILLMAN, D. B. 1954. The stratigraphy and structure of the Caplinger Mills Quadrangle, Missouri. Unpublished Master's thesis, University of Iowa.

MOORE, R. C., M. K. ELIAS, and N. D. NEWELL. 1936. A "Permian" flora from the Pennsylvanian of Kansas. Jour. Geol. 44:1–31.

———, and others. 1944. Correlation of Pennsylvanian formations of North America. Geol. Soc. Amer., Bull. 55:657–706.

———. 1948. Classification of Pennsylvanian rocks in Iowa, Missouri, Nebraska, and northern Oklahoma. Amer. Assoc. Petroleum Geologists, Bull. 32:2011–2040.

NEWBERRY, J. S. 1853. Structure and affinities of certain fossil plants of the Carboniferous era. Cleveland Ann. Sci. 1:157–162.

———. 1854. New species of fossil plants. Cleveland Ann. Sci. 2:2–3.

———. 1854. Fossil plants from the Ohio Coal Basin. Cleveland Ann. Sci. 106–108, 116–117, 152–153.

———. 1873. Description of fossil plants from the Coal Measures of Ohio. Ohio Geol. Surv., pt. 2, Palaeontol., sec. 3, 1:359–385.

NOÉ, A. C. 1923. The flora of the western Kentucky coal field. Kentucky Geol. Surv., ser. 6, 10:127–148.

———. 1925. Pennsylvanian flora of northern Illinois. Illinois Geol. Surv., Bull. 52. 113 p.

PIERCE, W. G., and W. H. COURTIER. 1937. Geology and coal resources of the southeastern Kansas coal field in Crawford, Cherokee, and Labette Counties. Kansas State Geol. Surv., Bull. 24:1–91.

READ, C. B. 1934. A flora of Pottsville age from the Mosquito Range, Colorado. U.S. Geol. Surv., Prof. Paper 185-D:79–91.

———, and C. W. MERRIAM. 1940. A Pennsylvanian flora from central Oregon. Amer. Jour. Sci. 238:107–111.

———. 1947. Pennsylvanian floral zones and floral provinces. Jour. Geol. 55:271–279.

———, and S. H. MAMAY. 1960. Upper Paleozoic floral zones of the United States. U. S. Geol. Surv., Prof. Paper 400-B: 381–383.

ROUND, E. M. 1926. Correlations of coal floras in Henry County, Missouri, and the Narragansett Basin. Bot. Gaz. 83: 61–69.

SEARIGHT, T. K. 1952. Geology of the Humansville Quadrangle, Missouri. Unpublished Master's thesis, University of Missouri.

SEARIGHT, W. V., and others. 1953. Classification of Desmoinesian (Pennsylvanian) of the northern Mid-Continent. Amer. Assoc. Petroleum Geologists, Bull. 37:2747–2749.

———. 1959. Pennsylvanian (Desmoinesian) of Missouri. Missouri Geol. Surv. and Water Resources, Rept. Inv. 25, 46 p.

SELLARDS, E. H. 1908. Fossil plants of the Upper Paleozoic of Kansas. Kansas University Geol. Surv. 9:386–480.

SEWARD, A. C. 1917. Fossil plants. vol. 3. 656 p. Cambridge, England: Cambridge University Press.

SHEPARD, F. P. 1959. Marine sediments. Sci. 130:141–149.

SHUTTS, C. F., and J. E. CANRIGHT. 1955. Recent collections of Pennsylvanian plant fossils in Indiana. Indiana Acad. Sci., Proc. 64:70–78.

SIEVER, R. 1956. Correlation chart of classification of the Pennsylvanian rocks of Illinois as of 1956. In H. R. Wanless, Classification of the Pennsylvanian Rocks of Illinois as of 1956. Illinois State Geol. Surv., Circ. 217, 14 p.

STEWART, W. N. 1950. Report on the Carr and Daniels collections of fossil plants from Mazon Creek. Illinois Acad. Sci., Trans. 43:41–45.

STOPES, M. C. 1914. The "Fern Ledges" Carboniferous flora of St. John, New Brunswick. Canada Geol. Surv., Mem. 41, 142 p.

UNGER, C. W. 1907. An account of the various contributions made to knowledge of the fossil flora of the southern anthracite coal field and the adjacent Paleozoic formations in Pennsylvania, with a list of the fossil plants. Schuylkill, County Historical Soc., Pub. 2:50–102.

WAGNER, R. H. 1962. A brief review of the stratigraphy and floral succession of the Carboniferous in NW. Spain. Quatrième Congrès pour l'avancement des Études de Stratigraphie et de Géologie du Carbonifère, Compte Rendu 3:753–762.

WALKER, R. A. 1961. Geology of the Filley Quadrangle, Cedar

County, Missouri. Unpublished Master's thesis, University of Missouri.

WALTON, R. H. 1953. An introduction to the study of fossil plants. 201 p. London: Adam and Charles Black.

WHITE, D. 1893. Flora of the outlying Carboniferous basins of southwestern Missouri. U.S. Geol. Surv., Bull. 98, 128 p.

———. 1897–98. Report on fossil plants from the McAlester coal field, Indian Territory, collected by Messers Taft and Richardson in 1879. U.S. Geol. Surv., 19th Ann. Rept., Econ. Geol. 3:457–534.

———. 1897. Age of the Lower Coals of Henry County, Missouri. Amer. Geol. Soc., Bull. 8:287–304.

———. 1899. Fossil flora of the Lower Coal Measures of Missouri. U.S. Geol. Surv., Monogr. 37:1–467.

———. 1900a. Relative ages of the Kanawha and Allegheny series as indicated by the fossil plants. Amer. Geol. Soc., Bull. 11:145–178.

———. 1900b. The stratigraphic succession of the fossil floras of the Pottsville Formation and the southern anthracite coal fields. U. S. Geol. Surv., 20th Ann. Rept. 2:755–930.

———. 1903. Summary of the fossil plants recorded from the Upper Carboniferous rocks and Permian formations of Kansas. In G. E. Adams, G. H. Girty and D. White, Stratigraphy and paleontology of the Upper Carboniferous rocks of the Kansas section. U. S. Geol. Surv., Bull. 211:85–117.

———. 1907. Report on fossil plants from the Coal Measures of Arkansas. U. S. Geol. Surv., Bull. 326:24–31.

———. 1909. The Upper Paleozoic Floras; their succession and range. Jour. Geol. 17:320–341.

———. 1911. Value of floral evidence in the marine strata as indicative of nearness of shores. Amer. Geol. Soc., Bull. 22:221–227.

———. 1913. The fossil flora of West Virginia. West Virginia Geol. Surv. 5 (A) part 2:390–453.

———. 1915. Notes on the fossil floras of the Pennsylvanian in Missouri. Missouri Bur. Geology and Mines, ser. 2, 13:256–262.

———. 1931. Climatic implications of the Pennsylvanian flora. Illinois State Geol. Surv., Bull. 60:271–281.

———. 1940. Fossil plants from the Stanley shale and Jackfork sandstone in southeastern Oklahoma and western Arkansas. U. S. Geol. Surv., Prof. Paper 186:43–66.

———. 1943. Lower Pennsylvanian species of *Mariopteris, Eremopteris, Diplothmema,* and *Aneimites* from the Appalachian region. U. S. Geol. Surv., Prof. Paper 197-C:85–140.

WINSLOW, A. 1891. A preliminary report on the coal deposits of Missouri. Missouri Geol. Surv. 226 p.

WOOD, H. C. 1860. Contributions to the Carboniferous floras of the United States. Philadelphia Acad. Nat. Sci., Proc. 12:236–240, 436–443, 519–522.

———. 1866. Contribution to the knowledge of the flora of the Coal Period in the United States. Amer. Philosoph. Soc., Trans. 13:341–349.

Wood, J. M. 1963. The Stanley Cemetery flora (Early Pennsylvanian) of Greene County, Indiana. Indiana Dept. Cons. Geol. Surv., Bull. 29, 73 p.

———, and J. E. Canright. 1954. The present status of paleobotany in Indiana with special reference to the fossils of Pennsylvanian age. Indiana Acad. Sci., Proc. 63:87–91.

Zeiller, R. 1886–1888. Bassin houiller de Valenciennes, description de la flore fossile. Études des gîtes min. de la France, 731 p.

Plates

The generic and species epithets are followed by the catalog number, the magnification, and the collection site.

Plate 1

Figure
1. *Asolanus camptotaenia* Wood; No. 1042; x 2.3; Clary Pit
2. *Asolanus camptotaenia* Wood; No. 1057; x 2.0; Clary Pit
3. *Aspidiaria* sp. Presl; No. 1076; x 1.0; Clary Pit
4. *Knorria* sp. Sternberg; No. 1037; x 1.8; Clary Pit

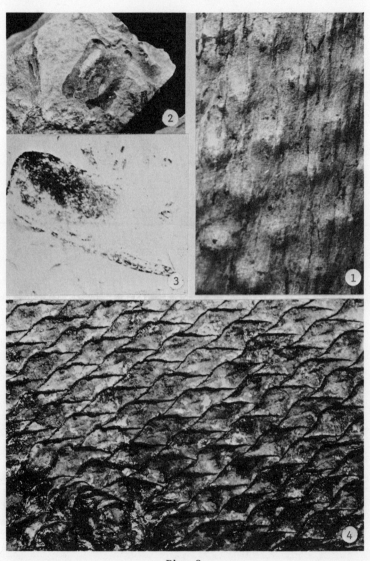

Plate 2

Figure
1. *Knorria* sp. Sternberg; No. 1038; x 1.7; Clary Pit
2. *Lepidocystis fraxiniformis* (Goepp.) Lesquereux; No. 1383; x 1.6; Gilkerson's Ford
3. *Lepidocystis fraxiniformis* (Goepp.) Lesquereux; No. 385; x 3.6; Clary Pit
4. *Lepidophloios van-ingeni* White; No. 1452; x 2.1; Gilkerson's Ford

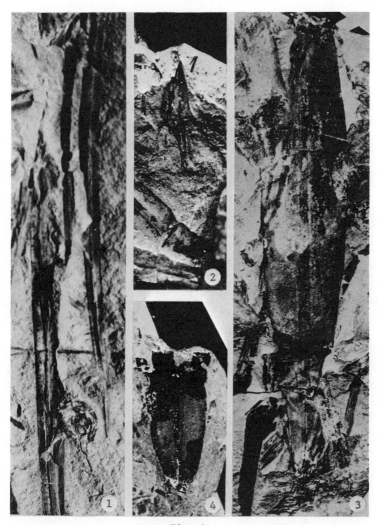

Plate 3

Figure
1. *Lepidophyllum* sp. Brongniart; No. 1357; x 1.7; Gilkerson's Ford
2. *Lepidostrobophyllum jenneyi* (White) Bell; No. 1434; x 3.0; Gilkerson's Ford
3. *Lepidostrobophyllum missouriense* (White) comb. nov.; No. 1406; x 1.7; Gilkerson's Ford
4. *Lepidostrobophyllum* sp. Hirmer; No. 1415; x 2.6; Gilkerson's Ford

Plate 4

Figure
1. *Lycopodites* sp. Brongniart; No. 1014; x 1.4; Clary Pit
2. *Annularia acicularis* (Dawson) White; No. 1588; x 2.1; Caplinger Mills
3. *Annularia sphenophylloides* (Zenker) Gutbier; No. 1700; x 2.4; Caplinger Mills

Plate 5

Figure
1. *Annularia galioides* (L & H) Kidston; No. 1680; x 2.3; Caplinger Mills
2. *Annularia stellata* (Schloth.) Wood; No. 406; x 3.0; Clary Pit

Plate 6

Figure
1. *Annularia stellata* (Schloth.) Wood; No. 372; x 1.7; Clary Pit
2. *Asterophyllites equisetiformis* (Schloth.) Brongniart; No. 572; x 2.8; Clary Pit
3. *Calamites carinatus* Sternberg; No. 1173; x 2.0; Hilty Mine
4. *Sphenophyllum emarginatum* Brongniart; No. 386; x 2.0; Clary Pit
5. *Sphenophyllum longifolium* Germar; No. 805; x 2.7; Clary Pit

Plate 7

Figure
1. *Calamites cisti* Brongniart; No. 1177; x 3.0; Hilty Mine
2. *Calamites cisti* Brongniart; No. 1171; x 2.0; Hilty Mine
3. *Calamites suckowi* Brongniart; No. 810; x 1.6; Clary Pit

154 Fossil Flora of the Drywood Formation

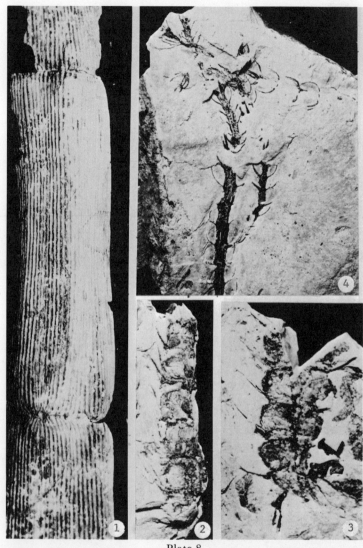

Plate 8

Figure
1. *Calamites cruciatus* Sternberg; No. 1081; x 1.0; Clary Pit
2. *Calamostachys paniculata* Weiss; No. 554; x 2.7; Clary Pit
3. *Calamostachys tuberculata* (Sternb.) Weiss; No. 667; x 2.7; Clary Pit
4. *Sphenophyllum fasciculatum* (Lx.) White; No. 863; x 2.8; Clary Pit

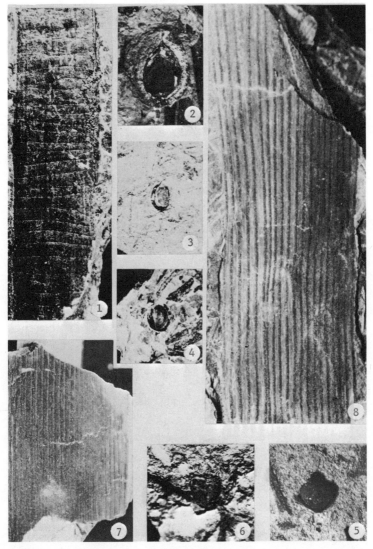

Plate 9

Figure
1. *Artisia transversa* (Artis) Sternberg; No. 1773; x 2.0; Creek west of Gilkerson's Ford
2. *Cardiocarpus late-alatum* Lesquereux; No. 1364; x 3.3; Gilkerson's Ford
3. *Cardiocarpus crassus* Lesquereux; No. 476; x 3; Clary Pit
4. *Cardiocarpus crassus* Lesquereux; No. 664; x 2.3; Clary Pit
5. *Cardiocarpus ovalis* Lesquereux; No. 1155; x 2.0; Hilty Mine
6. *Cardiocarpus ovalis* Lesquereux; No. 1510; x 2.0; Route J
7. *Cordaites crassinervis* Heer; No. 1739; x 2.1; Caplinger Mills
8. *Cordaites principalis* (Germar) Geinitz; No. 1678; x 2.3; Caplinger Mills

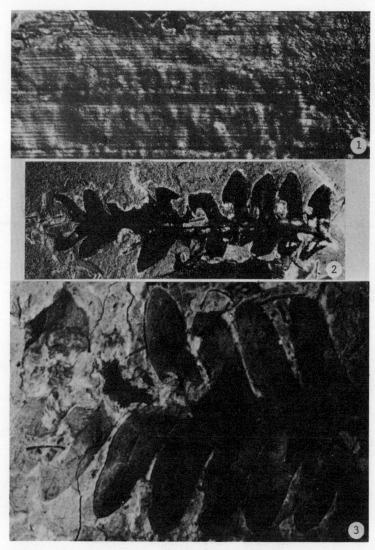

Plate 10

Figure
1. *Dorycordaites palmaeformis* (Goepp.) Zeiller; No. 861; x 3.1; Clary Pit
2. *Alethopteris davreuxi* (Brgt.) Goeppert; No. 1652; x 2.1; Caplinger Mills
3. *Alethopteris grandini* (Brgt.) Goeppert; No. 1638; x 2.0; Caplinger Mills

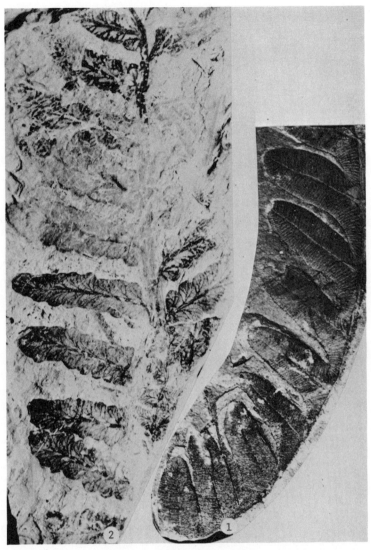

Plate 11

Figure
1. *Alethopteris serlii* (Brgt.) Goeppert; No. 1640; x 2.7; Caplinger Mills
2. *Alethopteris valida* Boulay; No. 1000; x 2.8; Clary Pit

Plate 12

Figure
1. *Alethopteris decurrens* (Artis) Zeiller; No. 824; x 2.9; Clary Pit
2. *Diplothmema furcatum* (Brgt.) Stur; No. 1573; x 1.8; Caplinger Mills
3. *Diplothmema obtusiloba* (Brgt.) Stur; No. 912; x 4.0; Clary Pit
4. *Eremopteris bilobata* White; No. 1525; x 1.3; Caplinger Mills

Plates 159

Plate 13

Figure
1. *Sphenopteris mixta* Schimper; No. 1410; x 2.9; Gilkerson's Ford
2. *Sphenopteris obtusiloba* ? Brongniart; No. 500; x 2.5; Clary Pit
3. *Sphenopteris* sp. Brongniart; No. 1055; x 2.3; Clary Pit

Plate 14

Figure
1. *Asterotheca* Presl sp. ?; No. 428; x 2.4; Clary Pit
2. *Mariopteris (Pseudopecopteris) decipiens* (Lx.) White; No. 1704; x 2.6; Caplinger Mills

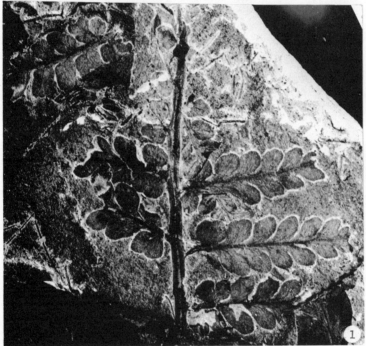

Plate 15

Figure
1. *Mariopteris (Pseudopecopteris) decipiens* (Lx.) White; No. 1716; x 2.0; Caplinger Mills
2. *Mariopteris speciosa* (Lx.) White; No. 1577; x 2.9; Caplinger Mills

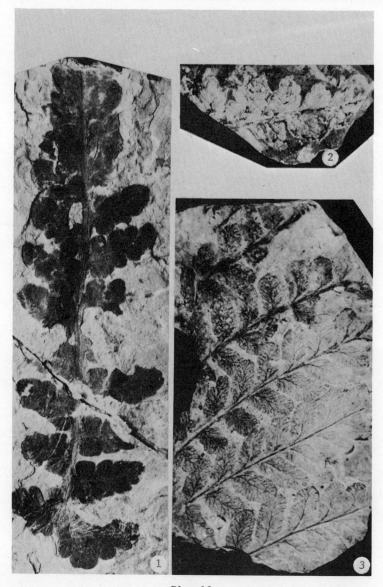

Plate 16

Figure
1. *Mariopteris speciosa* (Lx.) White; No. 1551; x 2.8; Caplinger Mills
2. *Mariopteris sphenopteroides* (Lx.) Zeiller; No. 911; x 2.7; Clary Pit
3. *Pecopteris clintoni* Lesquereux; No. 854; x 2.8; Clary Pit

Plates 163

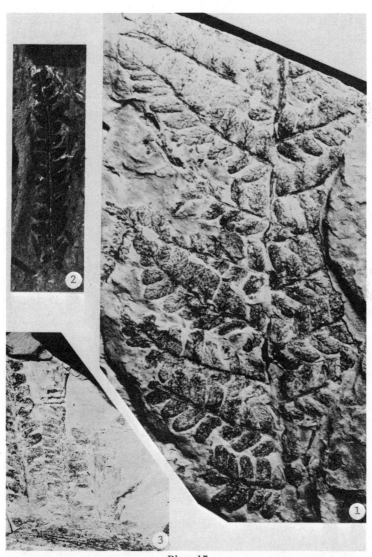

Plate 17

Figure
1. *Pecopteris clintoni* Lesquereux; No. 1020; x 3.0; Clary Pit
2. *Pecopteris dentata* Brongniart; No. 1399; x 2.3; Gilkerson's Ford
3. *Pecopteris pseudovestita* White; No. 1018; x 2.2; Clary Pit

Plate 18

Figure
1. *Pecopteris dentata* Brongniart; No. 807; x 1.4; Clary Pit
2. *Pecopteris pseudovestita* White; No. 1001; x 1.8; Clary Pit

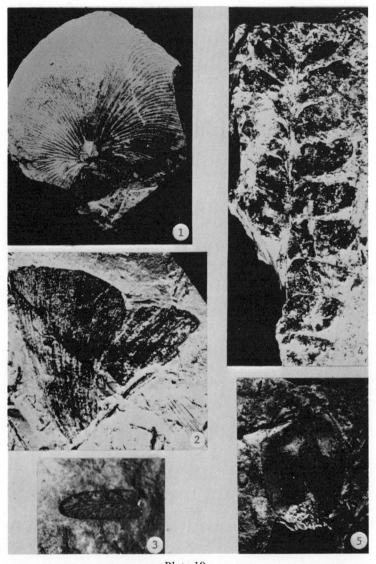

Plate 19

Figure
1. *Cyclopteris trichomanoides* Brongniart; No. 1765; x 1.7; Hilty Mine
2. *Cyclopteris trichomanoides* Brongniart; No. 753; x 2.1; Clary Pit
3. *Linopteris gilkersonensis* White; No. 1361; x 2.4; Gilkerson's Ford
4. *Neuropteris caudata* White; No. 904; x 2.7; Clary Pit
5. *Holcospermum* Nathorst sp. ?; No. 1705; x 2.1; Caplinger Mills

Plate 20

Figure
1. *Neuropteris elacerata* sp. nov. (Type Specimen); No. 913; x 3.4; Clary Pit
2. *Neuropteris heterophylla* Brongniart; No. 1665; x 2.2; Caplinger Mills

Plate 21

Figure
1. *Neuropteris heterophylla* Brongniart; No. 1635; x 2.1; Caplinger Mills
2. *Spiropteris* Schimper sp. ?; No. 1390; x 2.7; Gilkerson's Ford

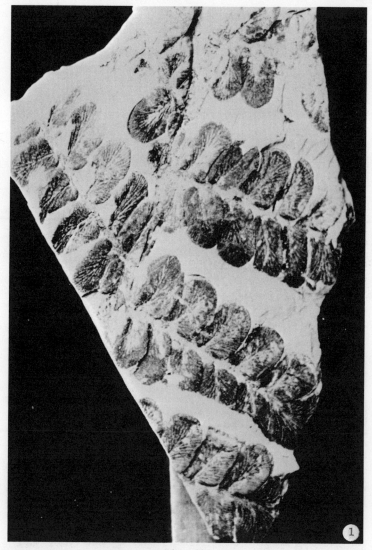

Plate 22

Figure
1. *Neuropteris missouriensis* Lesquereux; No. 757; x 3.1; Caplinger Mills

Plate 23

Figure
1. *Neuropteris ovata* Hoffm. forma *flexuosa* (Sternb.) Crookall; No. 1283; x 2.7; Hilty Mine
2. *Neuropteris ovata* Hoffm. forma *flexuosa* (Sternb.) Crookall; No. 1180; x 2.9; Hilty Mine

170 Fossil Flora of the Drywood Formation

Plate 24

Figure
1. *Neuropteris scheuchzeri* Hoffman; No. 1746; x 2.9; Caplinger Mills
   Cyclopteris sp. Brongniart (upper left-hand portion of plate)

About the Author

PHILIP W. BASSON has earned degrees from Eastern Michigan University (B.S. and Teacher's Certificate, 1959) and from the University of Missouri (M.A., 1961; Ph.D., 1965). He is a member of the Society of the Sigma Xi.

During his years of graduate study he taught in the Department of Botany at the University of Missouri, first as graduate teaching assistant and later as assistant instructor. In this period he also collected materials for the research described in this study and investigated new areas in southwestern Missouri as possible sources of additional materials.

Dr. Basson has served as Instructor of Botany in the Division of Biological Sciences, Section of Genetics, Development, and Physiology, at Cornell University and is, at present, Assistant Professor in Biology at the American University of Beirut.